A Lady's Visit
to the Gold Diggings
of Australia in 1852-53

ELLEN CLACY

CAMBRIDGE
UNIVERSITY PRESS

CAMBRIDGE UNIVERSITY PRESS

Cambridge, New York, Melbourne, Madrid, Cape Town,
Singapore, São Paolo, Delhi, Tokyo, Mexico City

Published in the United States of America by Cambridge University Press, New York

www.cambridge.org
Information on this title: www.cambridge.org/9781108039802

© in this compilation Cambridge University Press 2011

This edition first published 1853
This digitally printed version 2011

ISBN 978-1-108-03980-2 Paperback

CAMBRIDGE LIBRARY COLLECTION

Books of enduring scholarly value

Travel and Exploration

The history of travel writing dates back to the Bible, Caesar, the Vikings and the Crusaders, and its many themes include war, trade, science and recreation. Explorers from Columbus to Cook charted lands not previously visited by Western travellers, and were followed by merchants, missionaries, and colonists, who wrote accounts of their experiences. The development of steam power in the nineteenth century provided opportunities for increasing numbers of 'ordinary' people to travel further, more economically, and more safely, and resulted in great enthusiasm for travel writing among the reading public. Works included in this series range from first-hand descriptions of previously unrecorded places, to literary accounts of the strange habits of foreigners, to examples of the burgeoning numbers of guidebooks produced to satisfy the needs of a new kind of traveller - the tourist.

A Lady's Visit to the Gold Diggings of Australia in 1852-53

Mrs Charles (Ellen) Clacy (1830–1901) was a clergyman's daughter who, in 1852, travelled to the Australian goldfields. Published in 1853, on her return to England, this work, the first edition of which sold out almost immediately, is essentially a guide for prospective emigrants. It includes, within the lively narrative, practical advice on the cost of living, the labour market, gold-digging regulations, and marriage prospects. Mrs Clacy published several subsequent books, but her life remains obscure. Research suggests an illegitimate pregnancy or an absconding husband, unmentioned in the upbeat and respectable narrative, but possibly echoed by the highly coloured 'tale' of an anonymous emigrant woman, whose lover (twice) leaves her pregnant at the altar to go to the goldfields, with tragic consequences. However this relates to Mrs Clacy's actual circumstances, her writing vividly depicts the mixture of opportunity and hazard in nineteenth-century Australia, illuminating the country's early social history.

Cambridge University Press has long been a pioneer in the reissuing of out-of-print titles from its own backlist, producing digital reprints of books that are still sought after by scholars and students but could not be reprinted economically using traditional technology. The Cambridge Library Collection extends this activity to a wider range of books which are still of importance to researchers and professionals, either for the source material they contain, or as landmarks in the history of their academic discipline.

Drawing from the world-renowned collections in the Cambridge University Library, and guided by the advice of experts in each subject area, Cambridge University Press is using state-of-the-art scanning machines in its own Printing House to capture the content of each book selected for inclusion. The files are processed to give a consistently clear, crisp image, and the books finished to the high quality standard for which the Press is recognised around the world. The latest print-on-demand technology ensures that the books will remain available indefinitely, and that orders for single or multiple copies can quickly be supplied.

The Cambridge Library Collection will bring back to life books of enduring scholarly value (including out-of-copyright works originally issued by other publishers) across a wide range of disciplines in the humanities and social sciences and in science and technology.

A LADY'S VISIT

TO THE

GOLD DIGGINGS OF AUSTRALIA,

IN 1852-53.

Bendigo Creek.
Sin. 186.

London, Hurst & Blackett, 1853.

A LADY'S VISIT

TO THE

GOLD DIGGINGS OF AUSTRALIA,

IN 1852-53.

WRITTEN ON THE SPOT.

BY MRS. CHARLES CLACY.

LONDON:
HURST AND BLACKETT, PUBLISHERS,
SUCCESSORS TO HENRY COLBURN,
13, GREAT MARLBOROUGH STREET.
1853.

CONTENTS.

A LADY'S VISIT

TO THE

GOLD-DIGGINGS OF AUSTRALIA.

CHAPTER. I.

INTRODUCTORY REMARKS.

IT may be deemed presumptuous that one of my age and sex should venture to give to the public an account of personal adventures in a land which has so often been descanted upon by other and abler pens; but when I reflect on the many mothers, wives, and sisters in England, whose hearts are ever longing for information respecting the dangers and privations to which their relatives at the antipodes

B

are exposed, I cannot but hope that the pre-
sumption of my undertaking may be pardoned
in consideration of the pleasure which an
accurate description of some of the Australian Gold Fields may perhaps afford to
many; and although the time of my residence
in the colonies was short, I had the advantage
(not only in Melbourne, but whilst in the bush)
of constant intercourse with many experienced
diggers and old colonists—thus having every
facility for acquiring information respecting Victoria and the other colonies.

It was in the beginning of April, 185—,
that the excitement occasioned by the published accounts of the Victoria "Diggings,"
induced my brother to fling aside his Homer
and Euclid for the various "Guides" printed
for the benefit of the intending gold-seeker,
or to ponder over the shipping columns of the
daily papers. The love of adventure must be
contagious, for three weeks after (so rapid
were our preparations) found myself accom-
panying him to those auriferous regions. The
following pages will give an accurate detail
of my adventures there—in a lack of the
marvellous will consist their principal faults;

but not even to please would I venture to turn uninteresting truth into agreeable fiction. Of the few statistics which occur, I may safely say, as of the more personal portions, that they are strictly true.

CHAPTER II.

THE VOYAGE OUT.

EVERYTHING was ready—boxes packed, tinned, and corded; farewells taken, and ourselves whirling down by rail to Gravesend—too much excited—too full of the future—to experience that sickening of the heart, that desolation of the feelings, which usually accompanies an expatriation, however voluntary, from the dearly loved shores of one's native land. Although in the cloudy month of April, the sun shone brightly on the masts of our bonny bark, which lay in full sight of the windows of the "Old Falcon," where we had taken up our temporary quarters. The sea was very rough, but as we were anxious to get on board without farther delay, we entrusted our valuable lives in a four-oared boat, despite the dismal prognostications of our worthy host. A pleasant row that was, at one moment covered over with salt-water—the next riding on the top of a wave, ten times the size of our frail conveyance —then came a sudden concussion—in veering our rudder smashed into a smaller boat, which immediately filled and sank, and our rowers disheartened at this mishap would go no farther. The return was still rougher—my face smarted dreadfully from the cutting splashes of the salt-water ; they contrived, however, to land us safely at the "Old Falcon," though in a most pitiable plight ; charging only a sovereign for this delightful trip—very moderate, considering the number of salt-water baths they had given us gratis. In the evening a second trial proved more successful, and we reached our vessel safely.

A first night on board ship has in it something very strange, and the first awakening in the morning is still more so. To find oneself in a space of some six feet by eight, instead of a good-sized room, and lying in a cot, scarce

wide enough to turn round in, as a substitute
for a four-post bedstead, reminds you in no
very agreeable manner that you have exchanged
the comforts of Old England for the " roughing
it" of a sea life. The first sound that awoke
me was the " cheerily" song of the sailors,
as the anchor was heaved—not again, we
trusted, to be lowered till our eyes should rest
on the waters of Port Philip. And then the
cry of " raise tacks and sheets" (which I, in
nautical ignorance, interpreted " hay-stacks and
sheep") sent many a sluggard from their berths to
bid a last farewell to the banks of the Thames.

In the afternoon we parted company with
our steam-tug, and next morning, whilst off
the Isle of Wight, our pilot also took his departure. Sea-sickness now became the fashion,
but, as I cannot speak from experience of its
sensations, I shall altogether decline the subject.
On Friday, the 30th, we sighted Stark Point;
and as the last speck of English land faded
away in the distance, an intense feeling of
misery crept over me, as I reflected that perchance I had left those most dear to return
to them no more. But I forget; a description
of private feelings is, to uninterested readers, only

so much twaddle, besides being more egotist-
ical than even an account of personal adventures could extenuate; so, with the exception
of a few extracts from my "log," I shall jump
at once from the English Channel to the more
exciting shores of Victoria.

Wednesday, May 5, lat. 45° 57′ N., long.
11° 45′ W.—Whilst off the Bay of Biscay,
for the first time I had the pleasure of seeing
the phosphoric light in the water, and the
effect was indeed too beautiful to describe. I
gazed again and again, and, as the darkness
above became more dense, the silence of evening
more profound, and the moving lights beneath
more brilliant, I could have believed them the
eyes of the Undines, who had quitted their
cool grottos beneath the sea to gaze on the
daring ones who were sailing above them. At
times one of these stars of the ocean would
seem to linger around our vessel, as though
loth to leave the admiring eyes that watched
its glittering progress. * * * *

Sunday, 9, lat. 37° 53′ N., long. 15° 32′ W.
—Great excitement throughout the ship. Early
in the morning a homeward-bound sail hove
in sight, and as the sea was very calm, our

captain kindly promised to lower a boat and
send letters by her. What a scene then commenced; nothing but scribes and writing-desks
met the view, and nought was heard but the
scratching of pens, and energetic demands
for foreign letter-paper, vestas, or sealing-wax;
then came a rush on deck, to witness the important packet delivered to the care of the first
mate, and watch the progress of the little bark
that was to bear among so many homes the
glad tidings of our safety. On she came—
her stunsails set—her white sails glittering in
the sun—skimming like a sea-bird over the
waters. She proved to be the Maltese schooner
'Felix,' bound for Bremen. Her captain treated
the visitors from our ship with the greatest
politeness, promised to consign our letters to
the first pilot he should encounter off the
English coast, and sent his very last oranges
as a present to the ladies, for which we sincerely
thanked him; the increasing heat of the weather made them acceptable indeed.

Wednesday, 12, lat. 33° 19′ N., long. 17°
30 W —At about noon we sighted Madeira.
At first it appeared little more than a dark
cloud above the horizon; gradually the sides

of the rocks became clearly discernible, then the wind bore us onward, and soon all traces of the sunny isle were gone.

Friday, 28, lat. 4° 2′ N., long. 21° 30′ W. —Another opportunity of sending letters, but as this was the second time of so doing, the excitement was proportionately diminished. This vessel was bound for the port of Liverpool, from the coast of Africa; her cargo (so said those of our fellow-travellers who boarded her), consisted of ebony and gold-dust, her only passengers being monkeys and parrots.

Sunday, June 6, long. 24° 38′ W.—Crossed the Line, to the great satisfaction of all on board, as we had been becalmed more than a week, and were weary of gazing upon the unruffled waters around us, or watching the sails as they idly flapped to and fro. Chess, backgammon, books and cards, had ceased to beguile the hours away, and the only amusement left was lowering a boat and rowing about within a short distance of the ship, but this (even by those *not* pulling at the oars) was considered too fatiguing work, for a tropical sun was above us, and the heat was most intense. Our only resource was to give ourselves up to a sort of *dolce far niente* existence, and lounge upon the deck, sipping lemonade or lime-juice, beneath a large awning which extended from the fore to the mizen masts.

Tuesday, August 17, lat. 39° 28′ S., long. 136° 31′ E.—Early this morning one of the sailors died, and before noon the last services of the Church of England were read over his body; this was the first and only death that occurred during our long passage, and the solemnity of committing his last remains to their watery grave cast a saddening influence over the most thoughtless.   I shall never forget the moment when the sewn-up hammock, with a gaily-coloured flag wrapped round it, was launched into the deep; those who can witness with indifference a funeral on land, would, I think, find it impossible to resist the thrilling awe inspired by such an event at sea.

Friday, 20, lat. 38° 57′ S., long. 140° 8′ E. —Sighted Moonlight Head, the next day Cape Otway; and in the afternoon of Sunday, the 22nd, we entered the Heads, and our pilot came on board.   He was a smart, active fellow, and immediately anchored us within the bay (a heavy gale brewing); and then, after having done colonial justice to a substantial dinner, he edified us with the last Melbourne news. " Not a spare room or bed to be had— no living at all under a pound a-day—every one with ten fingers making ten to twenty pounds a-week." " Then of course no one goes to the diggings ?" " Oh, that pays better still —the gold obliged to be quarried—a pound weight of no value." The excitement that evening can scarcely be imagined, but it somewhat abated next morning on his telling us to diminish his accounts some 200 per cent.

Monday, 23.—The wind high, and blowing right against us. Compelled to remain at anchor, only too thankful to be in such safe quarters.

Tuesday, 24.—Got under weigh at half-past seven in the morning, and passed the wrecks of two vessels, whose captains had attempted to come in without a pilot, rather than wait for one—the increased number of vessels arriving, causing the pilots to be frequently all engaged. The bay, which is truly splendid, was crowded with shipping. In a few hours our anchor was lowered for the last time—boats were put off

towards our ship from Liardet's Beach—we
were lowered into the first that came alongside
—a twenty minutes' pull to the landing-place—
another minute, and we trod the golden shores
of Victoria.

CHAPTER III.

STAY IN MELBOURNE.

AT last we are in Australia. Our feet feel strange as they tread upon *terra firma,* and our *sea-legs* (to use a sailor's phrase) are not so ready to leave us after a four months' service, as we should have anticipated; but it matters little, for we are in the colonies, walking with undignified, awkward gait, not on a fashionable promenade, but upon a little wooden pier.

The first sounds that greet our ears are the noisy tones of some watermen, who are loitering on the building of wooden logs and boards, which we, as do the good people of Victoria, dignify with the undeserved title of *pier*. There they stand in their waterproof caps and skins— tolerably idle and exceedingly independent—with one eye on the look out for a fare, and the other cast longingly towards the open doors of Liardet's public-house, which is built a few yards from the landing-place, and alongside the main road to Melbourne.

"Ah, skipper! times isn't as they used to was," shouted one, addressing the captain of one of the vessels then lying in the bay, who was rowing himself to shore, with no other assistant or companion than a sailor-boy. The captain, a well-built, fine-looking specimen of an English seaman, merely laughed at this impromptu salutation.

"I say, skipper, I don't quite like that d—d stroke of yours."

No answer; but, as if completely deaf to these remarks, as well as the insulting tone in which they were delivered, the "skipper" continued giving his orders to his boy, and then leisurely ascended the steps. He walked straight up to the waterman, who was lounging against the railing.

"So, my fine fellow, you didn't quite admire

that stroke of mine. Now, I've another stroke that I think you'll admire still less," and with one blow he sent him reeling against the railing on the opposite side.

The waterman slowly recovered his equilibrium, muttering, "that was a safe dodge, as the gentleman knew he was the heaviest man of the two."

"Then never let your tongue say what your fist can't defend," was the cool retort, as another blow sent him staggering to his original place, amidst the unrestrained laughter of his companions, whilst the captain unconcernedly walked into Liardet's, whither we also betook ourselves, not a little surprised and amused by this our first introduction to colonial customs and manners.

The fact is, the watermen regard the masters of the ships in the bay as sworn enemies to their business; many are runaway sailors, and therefore, I suppose, have a natural antipathy that way; added to which, besides being no customers themselves, the "skippers," by the loan of their boats, often save their friends from the exorbitant charges these watermen levy.

Exorbitant they truly are. Not a boat would

they put off for the nearest ship in the bay
for less than a pound, and before I quitted
those regions, two and three times that sum
was often demanded for only one passenger.
We had just paid at the rate of only three
shillings and sixpence each, but this trifling
charge was in consideration of the large party
—more than a dozen—who had left our ship
in the same boat together.

Meanwhile we have entered Liardet's *en at-
tendant* the Melbourne omnibus, some of our
number, too impatient to wait longer, had
already started on foot. We were shown into a
clean, well-furnished sitting-room, with mahogany dining-table and chairs, and a showy glass
over the mantelpiece. An English-looking
barmaid entered, " Would the company like
some wine or spirits?" Some one ordered
sherry, of which I only remember that it was
vile trash at eight shillings a bottle.

And now the cry of " Here's the bus," brought
us quickly outside again, where we found several
new arrivals also waiting for it. I had hoped,
from the name, or rather misname, of the conveyance, to gladden my eyes with the sight of
something civilized. Alas, for my disappoint-

ment! There stood a long, tumble-to-pieces-looking waggon, not covered in, with a plank down each side to sit upon, and a miserable narrow plank it was. Into this vehicle were crammed a dozen people and an innumerable host of portmanteaus, large and small, carpet-bags, baskets, brown-paper parcels, bird-cage and inmate, &c., all of which, as is generally the case, were packed in a manner the most calculated to contribute the largest amount of inconvenience to the live portion of the cargo. And to drag this grand affair into Melbourne were harnessed thereto the most wretched-looking objects in the shape of horses that I had ever beheld.

A slight roll tells us we are off.

" And is *this* the beautiful scenery of Australia?" was my first melancholy reflection. Mud and swamp—swamp and mud—relieved here and there by some few trees which looked as starved and miserable as ourselves. The cattle we passed appeared in a wretched condition, and the human beings on the road seemed all to belong to one family, so truly Vandemonian was the cast of their countenances.

" The rainy season's not over," observed the
driver, in an apologetic tone. Our eyes and
uneasy limbs most *feelingly* corroborated his
statement, for as we moved along at a foot-pace,
the rolling of the omnibus, owing to the deep
ruts and heavy soil, brought us into most
unpleasant contact with the various packages
before-mentioned. On we went towards Melbourne—now stopping for the unhappy horses
to take breath—then passing our pedestrian
messmates, and now arriving at a small specimen of a swamp ; and whilst they (with trowsers
tucked high above the knee and boots well satu-
rated) step, slide and tumble manfully through
it, we give a fearful roll to the left, ditto, ditto
to the right, then a regular stand-still, or perhaps, by way of variety, are all but jolted over
the animals' heads, till at length all minor con-
siderations of bumps and bruises are merged
in the anxiety to escape without broken
bones.

" The Yarra," said the conductor. I looked
straight ahead, and innocently asked " Where ?"
for I could only discover a tract of marsh or
swamp, which I fancy must have resembled the
fens of Lincolnshire, as they were some years

ago, before draining was introduced into that county. Over Princes Bridge we now passed, up Swanston Street, then into Great Bourke Street, and now we stand opposite the Post-office — the appointed rendezvous with the walkers, who are there awaiting us. Splashed, wet and tired, and also, I must confess, very cross, right thankful was I to be carried over the dirty road and be safely deposited beneath the wooden portico outside the Post-office. Our ride to Melbourne cost us only half-a-crown a piece, and a shilling for every parcel. The distance we had come was between two and three miles.

The non-arrival of the mail-steamer left us now no other care save the all-important one of procuring food and shelter. Scouts were accordingly despatched to the best hotels; they returned with long faces—"full." The second-rate, and in fact every respectable inn and boarding or lodging-house were tried but with no better success. Here and there a solitary bed could be obtained, but for our digging party entire, which consisted of my brother, four shipmates, and myself, no accommodation could be procured, and we wished, if possible, to keep together. "It's a case," ejaculated one, casting his eyes to the slight roof above us as if calculating what sort of night shelter it would afford. At this moment the two last searchers approached, their countenances not quite so woebegone as before. "Well?" exclaimed we all in chorus, as we surrounded them, too impatient to interrogate at greater length. Thank Heavens! they had been successful! The housekeeper of a surgeon, who with his wife had just gone up to Forest Creek, would receive us to board and lodge for thirty shillings a week each; but as the accommodation was of the indifferent order, it was not as yet *une affaire arrangee.* On farther inquiry, we found the indifferent accommodation consisted in their being but one small sleeping-room for the gentlemen, and myself to share the bed and apartment of the temporary mistress. This was vastly superior to gipsying in the dirty streets, so we lost no time in securing our new berths, and ere very long, with appetites undiminished by these petty anxieties, we did ample justice to the dinner which our really kind hostess quickly placed before us.

The first night on shore after so long a

voyage could scarcely seem otherwise than strange, one missed the eternal rocking at which so many grumble on board ship. Dogs (Melbourne is full of them) kept up an incessant barking; revolvers were cracking in all directions until daybreak, giving one a pleasant idea of the state of society; and last, not least, of these annoyances was one unmentionable to ears polite, which would alone have sufficed to drive sleep away from poor wearied me. How I envied my companion, as accustomed to these disagreeables, she slept soundly by my side; but morning at length dawned, and I fell into a refreshing slumber.

The next few days were busy ones for all, though rather dismal to me, as I was confined almost entirely within doors, owing to the awful state of the streets; for in the colonies, at this season of the year, one may go out prepared for fine weather, with blue sky above, and dry under foot, and in less than an hour, should a *colonial* shower come on, be unable to cross some of the streets without a plank being placed from the middle of the road to the pathway, or the alternative of walking in water up to the knees.

This may seem a doleful and overdrawn picture of my first colonial experience, but we had arrived at a time when the colony presented its worst aspect to a stranger. The rainy season had been unusually protracted this year, in fact it was not yet considered entirely over, and the gold mines had completely upset everything and everybody, and put a stop to all improvements about the town or elsewhere.

Our party, on returning to the ship the day after our arrival, witnessed the French-leave-taking of all her crew, who during the absence of the captain, jumped overboard, and were quickly picked up and landed by the various boats about. This desertion of the ships by the sailors is an every-day occurrence; the diggings themselves, or the large amount they could obtain for the run home from another master, offer too many temptations. Consequently, our passengers had the amusement of hauling up from the hold their different goods and chattels; and so great was the confusion, that fully a week elapsed before they were all got to shore. Meanwhile we were getting initiated into colonial prices—money did indeed take to itself wings and fly away. Fire-arms were at a premium; one instance will suffice — my brother sold a six-barrelled revolver for which he had given sixty shillings at Baker's, in Fleet Street, for sixteen pounds, and the parting with it at that price was looked upon as a great favour. Imagine boots, and they very second-rate ones, at four pounds a pair. One of our between-deck passengers who had speculated with a small capital of forty pounds in boots and cutlery, told me afterwards that he had disposed of them the same evening he had landed, at a net profit of ninety pounds—no trifling addition to a poor man's purse. Labour was at a very high price, carpenters, boot and shoemakers, tailors, wheelwrights, joiners, smiths, glaziers, and, in fact, all useful trades, were earning from twenty to thirty shillings a day—the very men working on the roads could get eleven shillings per diem, and many a gentleman in this dis-arranged state of affairs, was glad to fling old habits aside and turn his hand to whatever came readiest. I knew one in particular, whose brother is at this moment serving as colonel in the army in India, a man more fitted for a gay London life than a residence in the colonies. The diggings were too dirty and uncivilized for his taste, his capital was quickly dwindling away

beneath the expenses of the comfortable life he
led at one of the best hotels in town, so he
turned to what as a boy he had learnt for
amusement, and obtained an addition to his
income of more than four hundred pounds a
year as house carpenter. In the morning you
might see him trudging off to his work, and
before night might meet him at some ball or
soirée among the élite of Melbourne.

I shall not attempt an elaborate description
of the town of Melbourne, or its neighbouring
villages. A subject so often and well discussed
might almost be omitted altogether. The town
is very well laid out; the streets (which are all
straight, running parallel with and across one
another) are very wide, but are incomplete, not
lighted, and many are unpaved. Owing to the
want of lamps, few, except when full moon,
dare stir out after dark. Some of the shops
are very fair; but the goods all partake too
largely of the flash order, for the purpose of
suiting the tastes of successful diggers, their
wives and families; it is ludicrous to see them
in the shops—men who, before the gold-mines
were discovered, toiled hard for their daily
bread, taking off half-a-dozen thick gold rings

from their fingers, and trying to pull on to their rough, well-hardened hands the best white kids, to be worn at some wedding party; whilst the wife, proud of the novel ornament, descants on the folly of hiding them beneath such useless articles as gloves.

The two principal streets are Collins Street and Elizabeth Street. The former runs east and west, the latter crossing it in the centre. Melbourne is built on two hills, and the view from the top of Collins Street East, is very striking on a fine day when well filled with passengers and vehicles. Down the eye passes till it reaches Elizabeth Street at the foot; then up again, and the moving mass seems like so many tiny black specks in the distance, and the country beyond looks but a little piece of green. A great deal of confusion arises from the want of their names being painted on the corners of the streets: to a stranger, this is particularly inconvenient, the more so, as being straight, they appear all alike on first acquaintance. The confusion is also increased by the same title, with slight variation, being applied to so many, as, for instance, Collins Street East; Collins Street West; Little Collins Street East; Little

C

Collins Street West, &c. &c.   Churches and
chapels for all sects and denominations meet
the eye; but the Established Church has, of
all, the worst provision for its members, only
two small churches being as yet completed;
and Sunday after Sunday do numbers return
from St. Peter's, unable to obtain even standing room beneath the porch.   For the gay,
there are two circuses and one theatre, where
the "ladies" who frequent it smoke short
tobacco-pipes in the boxes and dress-circle.

The country round is very pretty, particularly
Richmond and Collingwood; the latter will, I
expect, soon become part of Melbourne itself.
It is situated at the fashionable—that is, *east*—
end of Melbourne, and the buildings of the city
and this suburban village are making rapid
strides towards each other.   Of Richmond, I
may remark that it does possess a "Star and
Garter," though a very different affair to its
namesake at the antipodes, being only a small
public-house.   On the shores of the bay, at
nice driving distances, are Brighton and St.
Kilda.   Two or three fall-to-pieces bathing-
machines are at present the only stock in trade
of these watering-places; still, should some

would-be fashionables among my readers desire
to emigrate, it may gratify them to learn that
they need not forego the pleasure of visiting
Brighton in the season.

When I first arrived, as the weather was still
very cold and wet, my greatest source of dis-
comfort arose from the want of coal-fires, and
the draughts, which are innumerable, owing to
the slight manner in which the houses are run
up; in some the front entrance opens direct
into the sitting-rooms, very unpleasant, and
entirely precluding the "not at home" to an
unwelcome visitor. Wood fires have at best
but a cheerless look, and I often longed for the
bright blaze and merry fireside of an English
home. Firewood is sold at the rate of fifty
shillings for a good-sized barrow-full.

The colonists (I here speak of the old-estab-
lished ones) are naturally very hospitable, and
disposed to receive strangers with great kindness; but the present ferment has made them
forget everything in the glitter of their own
mines, and all comfort is laid aside; money is
the idol, and making it is the one mania which
absorbs every other thought.

The walking inhabitants are of themselves

a study: glance into the streets—all nations, classes, and costumes are represented there. Chinamen, with pigtails and loose trowsers; Aborigines, with a solitary blanket flung over them; Vandemonian pickpockets, with cunning eyes and light fingers—all, in truth, from the successful digger in his blue serge shirt, and with green veil still hanging round his wide-awake, to the fashionably-attired, newly-arrived "gent" from London, who stares around him in amazement and disgust. You may see, and hear too, some thoroughly colonial scenes in the streets. Once, in the middle of the day, when passing up Elizabeth Street, I heard the unmistakeable sound of a mob behind, and as it was gaining upon me, I turned into the enclosed ground in front of the Roman Catholic cathedral, to keep out of the way of the crowd. A man had been taken up for horse-stealing, and a rare ruffianly set of both sexes were following the prisoner and the two policemen who had him in charge. "If but six of ye were of my mind," shouted one, "it's this moment you'd release him." The crowd took the hint, and to it they set with right good will, yelling, swearing, and pushing, with awful violence. The owner of the stolen horse got up a counter demonstration, and every few yards, the procession was delayed by a trial of strength between the two parties. Ultimately the police conquered; but this is not always the case, and often lives are lost and limbs broken in the struggle, so weak is the force maintained by the colonial government for the preservation of order.

Another day, when passing the Post-office, a regular tropical shower of rain came on rather suddenly, and I hastened up to the platform for shelter. As I stood there, looking out into Great Bourke Street, a man and, I suppose, his wife passed by. He had a letter in his hand for the post; but as the pathway to the receiving-box looked very muddy, he made his companion take it to the box, whilst he himself, from beneath his umbrella, complacently watched her getting wet through. "Colonial politeness," thought I, as the happy couple walked on.

Sometimes a jovial wedding-party comes dashing through the streets; there they go, the bridegroom with one arm round his lady's waist, the other raising a champagne-bottle to his lips; the gay vehicles that follow contain company even more unrestrained, and from them noisier demonstrations of merriment may be heard. These diggers' weddings are all the rage, and bridal veils, white kid gloves, and, above all, orange blossoms are generally most difficult to procure at any price.

At times, you may see men, half-mad, throwing sovereigns, like halfpence, out of their pockets into the streets; and I once saw a digger, who was looking over a large quantity of bank-notes, deliberately tear to pieces and trample in the mud under his feet every soiled or ragged one he came to, swearing all the time at the gold-brokers for " giving him dirty paper money for pure Alexander gold; he wouldn't carry dirt in his pocket; not he; thank God! he'd plenty to tear up and spend too."

Melbourne is very full of Jews; on a Saturday, some of the streets are half closed. There are only two pawnbrokers in the town.

The most thriving trade there, is keeping an hotel or public-house, which always have a lamp before their doors. These at night serve as a beacon to the stranger to keep as far from them as possible, they being, with few exceptions, the resort, after dark, of the most ruffianly characters.

On the 2nd of September, the long-expected mail steamer arrived, and two days after we procured our letters from the Post-office. I may here remark, that the want of proper management in this department is the greatest cause of inconvenience to fresh arrivals, and to the inhabitants of Melbourne generally. There is but *one small window*, whence letters directed to lie at the office are given out; and as the ships from England daily discharged their living cargoes into Melbourne, the crowd round this inefficient delivering-place rendered getting one's letters the work, not of hours, but days. Newspapers, particularly pictorial ones, have, it would appear, a remarkable facility for being lost *en route*. Several numbers of the " Illustrated London News" had been sent me, and, although the letters posted with them arrived in safety, the papers themselves never made their appearance. I did hear that, when addressed to an uncolonial name, and with no grander direction than the Post-office itself, the clerks are apt to apropriate them—this is, perhaps, only a wee bit of Melbourne scandal.

The arrival of our letters from England left nothing now to detain us, and made us all anxious to commence our trip to the diggings, although the roads were in an awful condition. Still we would delay no longer, and the bustle of preparation began.   Stores of flour, tea, and sugar, tents and canvas, camp-ovens, cooking utensils, tin plates and pannikins, opossum rugs and blankets, drays, carts and horses, cradles, &c. &c., had to be looked at, bought and paid for.

On board ship, my brother had joined himself to a party of four young men, who had decided to give the diggings a trial.   Four other of our shipmates had also joined themselves into a digging-party, and when they heard of our intended departure, proposed travelling up together and separating on our arrival.   This was settled, and a proposal made that between the two sets they should raise funds to pur-chase a dray and horses, and make a speculation in flour, tea, &c., on which an immense profit was being made at the diggings.   It would also afford the convenience of taking up tents, cradles, and other articles impossible to carry up without. The dray cost one hundred pounds, and the two strong cart-horses ninety and one hundred pounds respectively. This, with the goods themselves, and a few sundries in the shape of harness and cords, made only a venture of about fifty pounds a-piece. While these arrangements were rapidly progressing, a few other parties wished to join ours for safety on the road, which was agreed to, and the day fixed upon for the departure was the 7th of September. Every one, except myself, was to walk, and we furthermore determined to " camp out" as much as possible, and thus avoid the vicinity of the inns and halting-places on the way, which are frequently the lurking-places of thieves and bushrangers.

———

On the Sunday previous to the day on which our journey was to commence, I had a little adventure, which pleased me at the time, though, but for the sequel, not worth mentioning here. I had walked with my brother and a friend to St. Peter's Church ; but we were a few minutes behind time, and therefore could find no unoccupied seat. Thus disappointed, we strolled over Princes Bridge on to the other side of the Yarra. Between the bridge and the beach, on the south side of the river, is a little city of tents, called Little Adelaide. They were inhabited by a number of families, that the rumour of the Victoria goldmines had induced to leave South Australia, and whose finances were unequal to the high prices in Melbourne.

Government levies a tax of five shillings a week on each tent, built upon land as wild and barren as the bleakest common in England.

We did not wander this morning towards Little Adelaide; but followed the Yarra in its winding course inland, in the direction of the Botanical Gardens.

Upon a gentle rise beside the river, not far enough away from Melbourne to be inconvenient, but yet sufficiently removed from its mud and noise, were pitched two tents, evidently new, with crimson paint still gay upon the round nobs of the centre posts, and looking altogether more in trim for a gala day in Merry England than a trip to the diggings. The sun

was high above our heads, and the day intensely
hot; so much so, that I could not resist the
temptation of tapping at the canvas door to ask
for a draught of water. A gentleman obeyed
the summons, and on learning the occasion of
this unceremonious visit, politely accommodated
me with a camp-stool and some delicious fresh
milk, in Melbourne almost a luxury. Whilst
I was imbibing this with no little relish, my
friends were entering into conversation with our
new acquaintance. The tents belonged to a
party just arrived by the steamer from England,
with everything complete for the diggings, to
which they meant to proceed in another week,
and where I had the pleasure of meeting them
again, though under different and very peculiar
circumstances. The tent which I had invaded
was inhabited by two, the elder of whom, a
powerfully-built man of thirty, formed a strong
contrast to his companion, a delicate-looking
youth, whose apparent age could not have exceeded sixteen years.

After a short rest, we returned to Melbourne,
well pleased with our little adventure.

The next day was hardly long enough for
our numerous preparations, and it was late be-

fore we retired to rest. Six was the hour
appointed for the next morning's breakfast.
Excited with anticipating the adventures to
commence on the morrow, no wonder that my
dreams should all be *golden* ones.

CHAPTER IV

CAMPING UP—MELBOURNE TO THE BLACK FOREST.

THE anxiously-expected morning at length commenced, and a dismal-looking morning it was—hazy and damp, with a small drizzling rain, which, from the gloomy aspect above, seemed likely to last. It was not, however, sufficient to damp our spirits, and the appointed hour found us all assembled to attack the last meal that we anticipated to make for some time to come beneath the shelter of a ceiling. At eight o'clock our united party was to start from the " Duke of York " hotel, and as that hour drew nigh, the unmistakeable signs of " something up," attracted a few idlers to witness our departure. In truth, we were a goodly party, and created no little sensation among the loungers—but I must regularly introduce our troop to my readers.

First then, I must mention two large drays, each drawn by a pair of stout horses—one the property of two Germans, who were bound for Forest Creek, the other belonged to ourselves and shipmates. There were three pack-horses—one (laden with a speculation in bran) belonged to a queer-looking sailor, who went by the name of Joe, the other two were under the care of a man named Gregory, who was going to rejoin his mates at Eagle Hawk Gully. As his destination was the farthest, and he was well acquainted with the roads, he ought to have been elected leader, but from some mis-management that dignity was conferred upon a stout old gentleman, who had taken a pleasure-trip to Mount Alexander, the previous summer.

Starting is almost always a tedious affair, nor was this particular case an exception. First one had forgotten something—another broke a strap, and a new one had to be pro-cured—then the dray was not properly packed,

and must be righted—some one else wanted
an extra "nobbler"—then a fresh, and still
a fresh delay, so that although eight was the
appointed hour, it was noon ere we bade
farewell to mine host of the "Duke of York."

At length the word of command was spoken.
Foremost came the gallant captain (as we had
dubbed him), and with him two ship doctors,
in partnership together, who carried the signs
of their profession along with them in the
shape of a most surgeon-like mahogany box.
Then came the two Germans, complacently
smoking their meerschaums, and attending to
their dray and horses, which latter, unlike their
masters, were of a very restless turn of mind.
After these came a party of six, among whom
was Gregory and two lively Frenchmen, who
kept up an incessant chattering. Joe walked
by himself, leading his pack-horse, then came
our four shipmates, two by two, and last, our
own particular five.

Most carried on their backs their individual
property—blankets, provisions for the road,
&c., rolled in a skin, and fastened over the
shoulders by leathern straps. This bundle goes
by the name of "swag," and is the digger's

usual accompaniment—it being too great a luxury to place upon a dray or pack-horse any-thing not absolutely necessary. This will be easily understood when it is known that carriers, during the winter, obtained £120 and some-times £150 a ton for conveying goods to Bendigo (about one hundred miles from Mel-bourne). Nor was the sum exorbitant, as besides the chance of a few weeks' stick in the mud, they run great risk of injuring their horses or bullocks; many a valuable beast has been obliged to be shot where it stood, it being found impossible to extricate it from the mud and swamp. At the time we started, the sum generally demanded was about £70 per ton. On the price of carriage up, depended of course the price of provisions at the diggings.

The weight of one of these " swags" is far from light; the provender for the road is itself by no means trifling, though that of course diminishes by the way, and lightens the load a little. Still there are the blankets, fire-arms, drinking and eating apparatus, clothing, chamois-leather for the gold that has yet to be dug, and numberless other cumbersome articles necessary for the digger. In every belt was stuck either a large knife or a tomahawk; two shouldered their guns (by the bye, rather imprudent, as the sight of fire-arms often brings down an attack); some had thick sticks, fit to fell a bullock; altogether, we seemed well prepared to encounter an entire army of bushrangers. I felt tolerably comfortable perched upon our dray, amid a mass of other soft lumber; a bag of flour formed an easy support to lean against; on either side I was well walled in by the canvas and poles of our tent; a large cheese made a convenient footstool. My attire, although well suited for the business on hand, would hardly have passed muster in any other situation. A dress of common dark blue serge, a felt wide-awake, and a waterproof coat wrapped round me, made a ludicrous assortment.

Going along at a foot-pace we descended Great Bourke Street, and made our first halt opposite the Post-office, where one of our party made a last effort to obtain a letter from his lady-love, which was, alas! unsuccessful. But we move on again—pass the Horse Bazaar—turn into Queen Street—up we go towards

Flemington, leaving the Melbourne cemetery on our right, and the flag-staff a little to the left; and now our journey may be considered fairly begun.

Just out of Melbourne, passing to the east of the Benevolent Asylum, we went over a little rise called Mount Pleasant, which, on a damp sort of a day, with the rain beating around one, seemed certainly a misnomer. After about two miles, we came to a branch-road leading to Pentridge, where the Government convict establishment is situated. This we left on our right, and through a line of country thickly wooded (consisting of red and white gum, stringy bark, cherry and other trees), we arrived at Flemington, which is about three miles and a half from town.

Flemington is a neat little village or township, consisting of about forty houses, a blacksmith's shop, several stores, and a good inn, built of brick and stone, with very fair accommodation for travellers, and a large stable and stock-yards.

After leaving Flemington, we passed several nice-looking homesteads; some are on a very large scale, and belong to gentlemen connected

with Melbourne, who prefer "living out of
town." On reaching the top of the hill
beyond Flemington there is a fine view of
Melbourne, the bay, William's Town, and the
surrounding country, but the miserable weather
prevented us at this time from properly en-
joying it. Sunshine was all we needed to
have made this portion of our travels truly
delightful.

The road was nicely level, fine trees
sheltered it on either side, whilst ever and
anon some rustic farm-house was passed, or
a coffee-shop, temporarily erected of canvas or
blankets, offered refreshment (such as it was),
and the latest news of the diggings to those
who had no objection to pay well for what
they had. This Flemington road (which is
considered the most pleasant in Victoria, or
at least anywhere near Melbourne) is very
good as far as Tulip Wright's, which we now
approached.

Wright's public-house is kept by the man
whose name it bears; it is a rambling ill-
built, but withal pleasing-looking edifice, built
chiefly of weather-board and shingle, with a
verandah all round. The whole is painted

white, and whilst at some distance from it
a passing ray of sunshine gave it a most
peculiar effect. In front of the principal
entrance is a thundering large lamp, a most
conspicuous looking object. Wright himself
was formerly in the police, and being a sharp
fellow, obtained the cognomen of "Tulip," by
which both he and his house have always been
known; and so inseparable have the names
become, that, whilst "Tulip Wright's" is
renowned well-nigh all over the colonies, the
simple name of the owner would create some
inquiries. The state of accommodation here
may be gathered from the success of some of
the party who had a *penchant* for "nobblers"
of brandy. "Nothing but bottled beer in the
house." "What could we have for dinner?"
inquired one, rather amused at this Hobson's
choice state of affairs. "The eatables was
only cold meat; and they couldn't cook
nothink fresh," was the curt reply. "Can
we sleep here?" "Yes—under your drays."
As we literally determined to "camp out" on
the journey, we passed on, without partaking
of their "cold eatables," or availing ourselves
of their permission to sleep under our own

drays, and, leaving the road to Sydney on our right, and the one to Keilor straight before us, we turned short off to the left towards the Deep Creek.

Of the two rejected routes I will give a very brief account.

The right-hand road leads to Sydney, *viâ* Kilmore, and many going to the diggings prefer using this road as far as that township. The country about here is very flat, stony and destitute of timber; occasionally the journey is varied by a water-hole or surface-spring. After several miles, a public-house called the " Lady of the Lake" is reached, which is reckoned by many the best country inn on this or any other road in the colonies. The accommodation is excellent, and the rooms well arranged, and independent of the house. There are ten or twelve rooms which, on a push, could accommodate fifty or sixty people; six are arranged in pairs for the convenience of married persons, and the fashionable trip during the honey-moon (particularly for diggers' weddings) is to the " Lady of the Lake." Whether Sir Walter's poem be the origin of the sign, or whether the swamps in the rear, I cannot say, but decidedly

there is no lake and no lady, though I have
heard of a buxom lass, the landlord's daughter,
who acts as barmaid, and is a great favourite.
This spot was the scene last May of a horrible
murder, which has added no little to the noto-
riety of the neighbourhood.

After several miles you at length arrive at
Kilmore, which is a large and thriving township,
containing two places of worship, several stores
and inns.   There is a resident magistrate with
his staff of officials, and a station for a detach-
ment of mounted police.   Kilmore is on the
main overland road from Melbourne to Sydney,
and, although not on the confines of the two
colonies, is rather an important place, from being
the last main township until you reach the
interior of New South Wales.   The Government buildings are commodious and well arranged.   There are several farms and stations
in the neighbourhood, but the country round
is flat and swampy.

The middle road leads you direct to Keilor,
and you must cross the Deep Creek in a dangerous part, as the banks thereabouts are very
steep, the stream (though narrow) very rapid,
and the bottom stony.   In 1851, the bridge (an

ordinary log one) was washed down by the floods, and for two months all communication was cut off. Government have now put a punt, which is worked backwards and forwards every half-hour from six in the morning till six at night, at certain fares, which are doubled after these hours. These fares are: for a passenger, 6*d.*; a horse or bullock, 1*s.*; a two-wheeled vehicle, 1*s.* 6*d.*; a loaded dray, 2*s.* The punt is tolerably well managed, except when the man gets intoxicated—not an unfrequent occurrence. When there was neither bridge nor punt, those who wished to cross were obliged to ford it; and so strong has been the current, that horses have been carried down one or two hundred yards before they could effect a landing. Keilor is a pretty little village with a good inn, several nice cottages, and a store or two. The country round is hilly and barren—scarcely any herbage and that little is rank and coarse; the timber is very scarce. This road to the diggings is not much used.

But to return to ourselves. The rain and bad roads made travelling so very wearisome, that before we had proceeded far it was unanimously agreed that we should halt and pitch

our first encampment. " Pitch our first encampment! how charming!" exclaims some
romantic reader, as though it were an easily
accomplished undertaking. Fixing a gipsy-
tent at a *fête champêtre*, with a smiling sky
above, and all requisites ready to hand, is one
thing, and attempting to sink poles and erect
tents out of blankets and rugs in a high
wind and pelting rain, is (if I may be allowed
the colonialism) "a horse of quite another
colour." Some sort of sheltering-places were
at length completed; the horses were taken
from the dray and tethered to some trees within
sight, and then we made preparations for
satisfying the unromantic cravings of hunger
—symptoms of which we all, more or less,
began to feel. With some difficulty a fire
was kindled and kept alight in the hollow
trunk of an old gum tree. A damper was
speedily made, which, with a plentiful supply
of steaks and boiled and roasted eggs, was a
supper by no means to be despised. The
eggs had been procured at four shillings a
dozen from a farm-house we had passed.

It was certainly the most curious tea-table
at which I had ever assisted. Chairs, of course,

there were none, we sat or lounged upon the
ground as best suited our tired limbs; tin
pannicans (holding about a pint) served as tea-
cups, and plates of the same metal in lieu of
china; a teapot was dispensed with; but a
portly substitute was there in the shape of an
immense iron kettle, just taken from the fire
and placed in the centre of our grand tea-ser-
vice, which being new, a lively imagination
might mistake for silver. Hot spirits, for those
desirous of imbibing them, followed our substantial repast; but fatigue and the dreary
weather had so completely damped all disposition to conviviality, that a very short space
of time found all fast asleep except the three
unfortunates on the watch, which was relieved
every two hours.

Wednesday, September 8.—I awoke rather
early this morning, not feeling over-comfortable
from having slept in my clothes all night,
which it is necessary to do on the journey, so
as never to be unprepared for any emergency.
A small corner of my brother's tent had been
partitioned off for my *bed-room;* it was quite
dark, so my first act on waking was to push

D

aside one of the blankets, still wet, which had
been my roof during the night, and thus admit
air and light into my apartments. Having
made my toilette—after a fashion—I joined
my companions on the watch, who were
deep in the mysteries of preparing something
eatable for breakfast. I discovered that their
efforts were concentrated on the formation of a
damper, which seemed to give them no little
difficulty. A damper is the legitimate, and, in
fact, only bread of the bush, and should be
made solely of flour and water, well mixed and
kneaded into a cake, as large as you like, but
not more than two inches in thickness, and
then placed among the hot ashes to bake. If
well-made, it is very sweet and a good substitute
for bread. The rain had, however, spoiled our
ashes, the dough would neither rise nor brown,
so in despair we mixed a fresh batch of flour
and water, and having fried some rashers of fat
bacon till they were nearly melted, we poured
the batter into the pan and let it fry till done.
This impromptu dish gave general satisfaction
and was pronounced a cross between a pancake
and a heavy suet pudding.

Breakfast over, our temporary residences were pulled down, the drays loaded, and our journey recommenced.

We soon reached the Deep Creek, and crossed by means of a punt, the charges being the same as the one at Keilor. Near here is a station belonging to Mr. Ryleigh, which is a happy specimen of a squatter's home — everything being managed in a superior manner. The house itself is erected on a rise and surrounded by an extensive garden, vinery and orchard, all well stocked and kept; some beautifully enclosed paddocks reach to the Creek, and give an English park-like appearance to the whole. The view from here over the bay and Brighton is splendid; you can almost distinguish Geelong. About a quarter of a mile off is a little hamlet with a neat Swiss-looking church, built over a school-room on a rise of ground; it has a most peculiar effect, and is the more singular as the economizing the ground could not be a consideration in the colony; on the left of the church is a pretty little parsonage, whitewashed, with slate roof and green-painted window-frames.

I still fancy, though our redoubtable captain

most strenuously denied it, that we had in some
manner gone out of our way; however that may
be, the roads seemed worse and worse as we
proceeded, and our pace became more tedious
as here and there it was up-hill work till at
length we reached the Keilor plains. It was
almost disheartening to look upon that vast
expanse of flat and dreary land except where
the eye lingered on the purple sides of Mount
Macedon, which rose far distant in front of us.
On entering the plains we passed two or three
little farm-houses, coffee-shops, &c., and encountered several parties coming home for a
trip to Melbourne. For ten miles we travelled
on dismally enough, for it rained a great deal,
and we were constantly obliged to halt to get
the horses rested a little. We now passed a
coffee-shop, which although only consisting of a
canvas tent and little wooden shed, has been
known to accommodate above forty people of a
night. As there are always plenty of bad characters lounging in the neighbourhood of such
places, we kept at a respectful distance, and did
not make our final halt till full two miles farther
on our road. Tents were again pitched, but
owing to their not being fastened over securely,

many of us got an unwished-for shower-bath during the night; but this is nothing—at the antipodes one soon learns to laugh at such trifles.

Thursday, 9.—This morning we were up betimes, some of our party being so sanguine as to anticipate making the " Bush Inn" before evening. As we proceeded, this hope quickly faded away. The Keilor plains seemed almost impassable, and what with pieces of rock here, and a water-hole there, crossing them was more dangerous than agreeable. Now one passed a broken-down dray ; then one's ears were horrified at the oaths an unhappy wight was venting at a mud-hole into which he had stumbled. A comical object he looked, as, half-seas-over, he attempted to pull on a mud-covered boot, which he had just extricated from the hole where it and his leg had parted company. A piece of wood, which his imagination transformed into a shoe-horn, was in his hand. " Put it into the larboard side," (suiting the action to the word), " there it goes—damn her, she won't come on ! Put it into the starboard side—there it goes—well done, old girl," and he triumphantly rose from the ground, and reeled away.

With a hearty laugh, we proceeded on our road, and after passing two or three coffee-tents, we arrived at Gregory's Inn. The landlord is considered the best on the road, and is a practical example of what honesty and industry may achieve. He commenced some nine months before without a shilling—his tarpaulin tent and small stock of tea, sugar, coffee, &c., being a loan. He has now a large weather-board house, capable of making up one hundred beds, and even then unable to accommodate all his visitors, so numerous are they, from the good name he bears. Here we got a capital cold dinner of meat, bread, cheese, coffee, tea, &c., for three shillings a-piece, and, somewhat refreshed, went forwards in better spirits, though the accounts we heard there of the bad roads in the Black Forest would have disheartened many.

Mount Macedon now formed quite a beautiful object on our right: a little below that mountain appeared a smaller one, called the Bald Hill, from its peak being quite barren, and the soil of a white limestone and quartzy nature, which gives it a most peculiar and splendid appearance when the sun's rays are shining

upon it. As we advanced, the thickly-wooded
sides of Mount Macedon became more distinct,
and our proximity to a part of the country
which we knew to be auriferous, exercised an
unaccountable yet pleasureable influence over
our spirits, which was perhaps increased by the
loveliness of the spot where we now pitched
our tents for the evening. It was at the foot
of the Gap. The stately gum-tree, the shea-
oak, with its gracefully drooping foliage, the
perfumed yellow blossom of the mimosa, the
richly-wooded mountain in the background,
united to form a picture too magnificent to
describe. The ground was carpeted with wild
flowers ; the sarsaparilla blossoms creeping
everywhere ; before us slowly rippled a clear
streamlet, reflecting a thousand times the
deepening tints which the last rays of the
setting sun flung over the surrounding scenery ;
the air rang with the cawing of the numerous
cockatoos and parrots of all hues and colours
who made the woods resound with their tones,
whilst their restless movements and gay plumage
gave life and piquancy to the scene.

This night our beds were composed of the
mimosa, which has a perfume like the hawthorn.

The softest-looking branches were selected, cut down, and flung upon the ground beneath the tents, and formed a bed which, to my wearied limbs, appeared the softest and most luxuriant upon which I had slept since my arrival in the colonies.

Friday, 10.—With some reluctance 1 aroused myself from a very heavy slumber produced by the over fatigue of the preceding day. I found every one preparing to start ; kindly considerate, my companions thought a good sleep more refreshing for me than breakfast, and had deferred awakening me till quite obliged, so taking a few sailors' biscuits in my pocket to munch on the way, I bade farewell to a spot whose natural beauties I have never seen surpassed.

Proceeding onwards, we skirted thc Bald Hill, and entering rather a scrubby tract, crossed a creek more awkward for our drays than dangerous to ourselves ; we then passed two or three little coffee-shops, which being tents are always shifting their quarters, crossed another plain, very stony and in places swampy, which terminated in a thickly-wooded tract of gum and wattle trees. Into this wood we now entered. After about five miles uncomfortable travelling we reached the "Bush Inn."

I must here observe that no *distinct* road is ever cut out, but the whole country is cut up into innumerable tracks by the carts and drays, and which are awfully bewildering to the new-comer as they run here and there, now crossing a swamp, now a rocky place, here a creek, there a hillock, and yet, in many cases, all leading *bonâ fide* to the same place.

The "Bush Inn" (the genuine one, for there are two) consists of a large, well-built, brick and weather-board house, with bed-rooms for private families. There is a detached weatherboard, and stone kitchen, and tap-room, with sleeping-lofts above, a large yard with sheds and good stabling. A portion of the house and stables is always engaged for the use of the escort. About two hundred yards off is the "New Bush Inn," somewhat similar to the other, not quite so large, with an attempt at a garden. The charges at these houses are enormous. Five and six shillings per meal, seven-and-sixpence for a bottle of ale, and one shilling for half a glass or "nobbler" of brandy.

About half a mile distant is a large station belonging to Mr. Watson; the houses, huts and yards are very prettily laid out, and, in a few years he will have the finest vineyard in the neighbourhood. Two miles to the east is the residence of Mr. Poullett, Commissioner of Crown Lands, which is very pleasantly situated on the banks of an ever-running stream. The paddock, which is a large one (10 square miles, or 6400 acres), is well wooded. Some new police barracks and stabling yards are in the course of erection.

We did not linger in the "Bush Inn," but pursued our way over a marshy flat, crossed a dangerous creek, and having ascended a steep and thickly wooded hill on the skirts of the Black Forest, we halted and pitched our tents. It was little more than mid-day, but the road had been fearful—as bad as wading through a mire; men and beasts were worn out, and it was thought advisable to recruit well before entering the dreaded precincts of the Black Forest. Fires were lit, supper was cooked, spirits and pipes made their appearance, songs were sung, and a few of the awful exploits of Black Douglas and his followers were related. Later in the evening, an opossum was shot by one of us. Its skin was very soft, with rich, brown hair.

Saturday, 11.—A dismal wet day—we remained stationary, as many of our party were still foot-sore, and all were glad of a rest. Some went out shooting, but returned with only a few parrots and cockatoos, which they roasted, and pronounced nice eating. Towards evening, a party of four, returning from the diggings, encamped at a little distance from us. Some of our loiterers made their acquaintance. They had passed the previous night in the Black Forest, having wandered out of their way. To add to their misfortunes, they had been attacked by three well-armed bushrangers, whom they had compelled to desist from their attempt, not, however, before two of the poor men had been wounded, one rather severely. Hardly had they recovered this shock, than they were horrified by the sudden discovery in a sequestered spot of some human bones, strewn upon the ground beside a broken-down cart. Whether accident or design had brought these unfortunates to an untimely end, none know ; but this ominous appearance seemed to have terrified them even more than the bushrangers themselves. These accounts sobered our party not a little, and it was deemed advisable to double the watch that night.

CHAPTER V.

CAMPING UP—BLACK FOREST TO EAGLE HAWK GULLY.

Sunday, 12.—A LOVELY summer morning, which raised our spirits to something like their usual tone, with the exception of our gallant (?) captain, who resigned his post, declaring it his intention to return to Melbourne with the four returning diggers. Poor fellow! their awful account of the Black Forest had been too much for his courage. Gregory was elected in his place, and wishing him a pleasant trip home, our journey was resumed as usual, and we entered the forest. Here the trees grow very closely together; in some places they are so thickly set that the rear-guard of the escort cannot see the advance-guard in the march. There is a slight undergrowth of scrub. We saw some of the choicest of the *Erica* tribe in full bloom, like a beautiful crimson waxen bell-blossom, and once whilst walking (which I frequently did to relieve the monotony of being perched on the dray by myself) I saw a fine specimen of the *Oreludiæ* at the foot of a tree growing from the wood ; it was something like a yellow sweet-pea, but really too beautiful to describe. The barks of the trees, and also the ground, have a black, charred appearance (hence the name of the forest) ; this is said to have been caused by its having once been on fire. Many of the ambuscades of the noted Douglas were passed, and the scenes of some most fearful murders pointed out. We only halted once—so anxious were we to leave behind us this dreaded spot—and at sunset reached the borders of the Five Mile Creek.

Monday, 13.—Another fine day. Crossed the Five Mile Creek by means of a rickety sort of bridge. There are two inns here, with plenty of accommodation for man and beast. We patronized neither, but made the best of

our way towards Kyneton. Our road lay
through a densely wooded country till we
arrived at Jacomb's Station; this we left, and
turning to the right, soon reached Kyneton,
which lies on the river Campaspe.

Carlshrue lies to the right, about three miles
distant, on rather low land; this is the chief
station of the Government escort; the barrack
accommodation is first-rate, with stabling and
paddocks for the horses, &c.

Kyneton is about sixty-one miles from Melbourne. There are two large inns, with ample
accommodation for four hundred people between
them, several stores, with almost every needful
article. A neat little church, capable of holding
nearly three hundred persons, with a school and
parsonage. There is a resident magistrate and
constabulary, with a police-court and gaol in
progress of erection. The township is rather
straggling, but what houses there are have a
very picturesque appearance. The only draw-
back to this little town is the badness of the
streets. Although it is rather on an elevated
spot, the streets and roads, from the loamy
nature of the soil, are a perfect quagmire, even
abominable in summer time. The charges here

are high, but not extortionate, as, besides the two inns alluded to, there are several coffee-shops and lodging-houses; so competition has its effect even in the bush.

The Campaspe is a large river, and is crossed by a substantial timber bridge.

We still adhered to our original plan of camping out; a few necessaries were purchased in the town, and after continuing our journey to a little distance from it, we halted for the night.

Tuesday 14.—This morning commenced with a colonial shower, which gave us all a good drenching. Started about eight o'clock; returned to Kyneton; crossed the bridge, and passed several farm-houses. The country here is very changeable, sometimes flat and boggy, at others, very hilly and stony. We were obliged to ford several small creeks, evidently tributaries to the Campaspe, and at about ten miles from Kyneton, entered the Coliban range, which is thickly wooded. The river itself is about fourteen miles from Kyneton. Here we camped in the pouring rain. Some of our party walked to the town of Malmsbury, about a mile and a half from our camping place. The town

consisted of about three tents, and an inn dig-
nified by the appellation of the "Malmsbury
Hotel." It is a two-storied, weather-board, and
pale house, painted blue, with a lamp before
it of many colours, large enough for half-a-dozen
people to dine in. It (the inn, not the lamp,)
is capable of accommodating two hundred people,
independent of which there is a large tent,
similar to the booths at a fair, about 100 feet
long by 30 wide, for the convenience of those
who prefer sleeping under cover when the house
is full. Being hungry with their walk, our
comrades dined here, for which they paid
3s. 6d. a-piece; ale was 1s. 6d. a glass; brandy
2s. per half glass, or "nobbler;" cheese, 4s. 6d.
a pound; bread, 5s. the four-pound loaf; wine,
25s. a bottle. By the time they returned, we had
struck our tents, intending to cross a muddy-
banked creek that lay in our road that evening,
as we were told that the waters might be too
swollen to do it next day. The water reached
above their waists, and as my usual post was
very insecure, I was obliged to be carried over
on their shoulders, which did not prevent my
feet from being thoroughly soaked before reaching the other side, where we remained all night.

Wednesday, 15.—Rainy day again, so much so, that we thought it advisable not to shift our quarters. In the afternoon, three returning diggers pitched their tents not far from ours. They were rather sociable, and gave us a good account of the diggings. They had themselves been very fortunate. On the same day that we had been idly resting on the borders of the Black Forest, they had succeeded in taking twenty-three pounds weight out of their claim, and two days after, two hundred and six ounces more, making, in all, gold to the value (in England) of about eighteen hundred pounds. They were returning to Melbourne for a spree, (which means to fling their gains away as quickly as possible,) and then as soon as the dry season was regularly set in, they meant to return to Bendigo for another spell at work. On representing to them the folly of not making better use of their hard-earned wages, the answer invariably was, " Plenty more to be got where this came from," an apt illustration of the proverb, " light come, light go." Two of these diggers had with them their licences for the current month, which they offered to sell for ten shillings each; two of our company purchased them. This, although a common proceeding, was quite illegal, and, of course, the two purchasers had to assume for the rest of the month the names of the parties to whom the licences had been issued. As evening approached, our new acquaintances became very sociable, and amused us with their account of the diggings; and the subject of licensing being naturally discussed, led to our being initiated into the various means of evading it, and the penalties incurred thereby. One story they related amused us at the time, and as it is true I will repeat it here, though I fancy the lack of oral communication will subtract from it what little interest it did possess.

Before I commence, I must give my readers some little insight into the nature of the licence tax itself. The licence, (for which thirty shillings, or half an ounce of gold, is paid per month) is in the following form:

VICTORIA GOLD LICENCE.

No. 1710,   Sept. 3, 1852.

The Bearer, Henry Clements, having paid to me the Sum of One Pound, Ten Shillings, on account of the Territorial Revenue, I hereby

Licenee him to dig, search for, and remove Gold on and from any such Crown Land within the Upper Lodden District, as I shall assign to him for that purpose during the month of September, 1852, not within half-a-mile of any Head station.

This Licence is not transferable, and to be produced whenever demanded by me or any other person acting under the Authority of the Government, and to be returned when another Licence is issued.

           (*Signed*)   B. BAXTER, Commissioner.

At the back of the Licence are the following rules :

Regulations to be observed by the Persons digging for Gold, or otherwise employed at the Gold Fields.

1. Every Licensed Person must always have his Licence with him, ready to be produced whenever demanded by a Commissioner, or Person acting under his instructions, otherwise he is liable to be proceeded against as an Unlicensed person.

2. Every Person digging for Gold, or occu-pying Land, without a Licence, is liable by Law to be fined, for the first offence, not exceeding £5; for a second offence, not exceeding £15; and for a subsequent offence, not exceeding £30.

3. Digging for Gold is not allowed within Ten feet of any Public Road, nor are the Roads to be undermined.

4. Tents or buildings are not to be erected within Twenty feet of each other, or within Twenty feet of any Creek.

5. It is enjoined that all Persons at the Gold Fields maintain and assist in maintaining a due and proper observance of Sundays.

————

So great is the crowd around the Commis-sioner's tent at the beginning of the month, that it is a matter of difficulty to procure it, and consequently the inspectors rarely begin their rounds before the 10th, when (as they generally vary the fine according to the date at which the delinquency is discovered), a non-licensed digger would have the pleasure of ac-companying a crowd of similar offenders to the

Commissioners, sometimes four or five miles
from his working-place, pay a fine of about £3,
and take out a licence. After the 20th of the
month, the fine inflicted is generally from £5 to
£10 and a licence, which is rather a dear price
to pay for a few days' permission to dig, as a
licence, although granted on the 30th of one
month, would be unavailable for the next. The
inspectors are generally strong-built, rough-
looking customers, they dress like the generality
of the diggers, and are only known by their
carrying a gun in lieu of a pick or shovel.
Delinquents unable to pay the fine, have the
pleasure of working it out on the roads.

Now for my story—such as it is.

Mike and Robert were two as good mates as
any at the Mount Alexander diggings. They
had had a good spell of hard work, and, as is
usually the way, returned to Melbourne for a
holiday at Christmas-time; and then it was
that the bright eyes of Susan Hinton first sowed
discord between them. Mike was the successful
wooer, and the old man gave his consent; for
Mike, with one exception, had contrived to
make himself a favourite with both father and
daughter. The exception was this. Old Hin-

ton was a strict disciplinarian—one of what is
called the " good old school"—he hated radicals,
revolutionists, and reformers, or any opposition
to Church or State. Mike, on the contrary,
loved nothing better than to hold forth against
the powers that be; and it was his greatest
boast that Government had never pocketed a
farthing from him in the way of a licence.
This, in the old man's eyes, was his solitary
fault, and when Mike declared his intention of
taking another trip to the "lottery fields" before
taking a ticket in the even greater lottery of
marriage, he solemnly declared that no daughter
of his should ever marry a man who had been
openly convicted of in any way evading the
licence fee.

This declaration from any other man, who
had already promised his daughter in marriage,
would not have had much weight; but Mike
knew the stern, strict character of Hinton, and
respected this determination accordingly. The
day of their departure arrived, and with a tearful
injunction to bear in mind her father's wishes,
Susan bade her lover farewell, and Robert and
he proceeded on their journey. Full of his own
happiness, Mike had never suspected his com-

rade's love for Susan, and little dreamt he of
the hatred against himself to which it had
given birth—hatred the more to be dreaded
since it was concealed under a most friendly
exterior.

For the first month Mike behaved to the
very letter of the law, and having for the sum
of £1 10s. purchased his legal right to dig for
gold, felt himself a most exemplary character.
Success again crowned their efforts, and a
speedy return to Melbourne was contemplated.
In the ardour of this exciting work another
month commenced, and Mike at first forgot
and then neglected to renew his licence. "The
inspector rarely came his rounds before the
14th; the neighbourhood was considered deserted—fairly 'worked out;' he'd never come
round there." Thus argued Mike, and his
friend cordially agreed with him. "Lose a
day's work standing outside the Commissioner's
tent broiling in a crowd, when two days would
finish the job? Not he, indeed! Mike might
please himself, but *he* shouldn't get a licence;"
and this determination on the part of his
"mate" settled the matter.

In one respect Mike's self-security was not

unfounded; the gully in which their tent was
now pitched was nearly deserted. Some while
previous there had been a great rush to the
place, so great that it was almost excavated;
then the rush took a different direction, and few
now cared to work on the two or three spots
that had been left untouched. Like many other
localities considered "worked out," as much
remained in the ground as had been taken from
it, and as each day added to their store, Mike's
hilarity increased.

It was now the 10th of the month; their
hole had been fairly "bottomed," a nice little
nest of nuggets discovered, their gains divided,
and the gold sent down to the escort-office for
transit to Melbourne. A few buckets-full of
good washing-stuff was all that was left undone.
"To-day will finish that," thought Mike, and
to it he set with hearty good-will, to the intense
satisfaction of his comrade, who sat watching
him at a little distance. Suddenly Mike felt
a heavy hand upon his shoulder: he looked up,
and saw before him—the inspector. He had
already with him a large body of defaulters, and
Mike little doubted but that he must be added
to their number. Old Hinton's determined

E

speech, Susan's parting words and tears, flashed across his mind.

" You've lost your bonnie bride," muttered Robert, loud enough to reach his rival's ears.

Mike glanced at him, and the look of triumph he saw there roused every spark of energy within him, and it was in a tone of well-assumed composure that he replied to the inspector, "My licence is in my pocket, and my coat is below there ;" and without a moment's hesitation sprang into his hole to fetch it. Some minutes elapsed. The inspector waxed impatient. A suspicion of the truth flashed across Robert's mind, and he too descended the hole. *There* was the coat and the licence of the past month in the pocket ; but the owner had gone, vanished, and an excavation on one side which led into the next hole and thence into a complete labyrinth underground, plainly pointed out the method of escape. Seeing no use in ferreting the delinquent out of so dangerous a place, the inspector sulkily withdrew, though not without venting some of his ill-humour upon Robert, at whose representations, made to him the day previous, he had come so far out of his road.

But let us return to Mike. By a happy thought, he had suddenly remembered that whilst working some days before in the hole, his pick had let in daylight on one side, and the desperate hope presented itself to his mind that he might make a passage into the next pit, which he knew led into others, and thus escape. His success was beyond his expectation; and he regained the open air at a sufficient distance from his late quarters to escape observation. Once able to reflect calmly upon the event of the morning, it required little discrimination to fix upon Robert his real share in it. And now there was no time to lose in returning to Melbourne, and prevent by a speedy marriage any further attempt to set his intended father-in-law against him. The roads were dry, for it was the sultry month of February; and two days saw him beside his lady-love.

Although railroads are as yet unknown in Australia, everything goes on at railroad speed; and a marriage concocted one day is frequently solemnized the next. His eagerness, therefore, was no way remarkable. No time was lost; and when, three days after Mike's return, Robert (with his head full of plots and machinations) presented himself at old Hinton's door, he found them all at a well-spread wedding breakfast, round which were gathered a merry party, listening with a digger's interest to the way in which the happy bridegroom had evaded the inspector. Mike had wisely kept the story till Susan was his wife.

Thursday 16. — With great delight we hailed the prognostications of a fine day, and, after having eaten a hearty breakfast on the strength of it, we recommenced our travels, and crossed the Coliban Bridge. The Coliban is a fine river running through a beautiful valley bounded with green trees; the bridge is a timber one, out of repair, and dangerous. A township called Malmsbury has been laid out here in small allotments with the expectation of a future city; but as yet not a house has been erected, with the exception of the "hotel" before mentioned, putting one in mind of the American Eden in "Martin Chuzzlewit." A mile beyond the Coliban are the washing huts of John Orr's Station, and about three miles to the left is his residence; the house is stone, with verandahs, the garden and vineyards are prettily laid out.

After passing the bridge, we took the right-hand road, which led us through a low country, and across two or three tributary creeks; we then reached the neighbourhood of Saw-pit Gully, so called from the number of saw-pits there, which formerly gave employment to numerous sawyers, whose occupation—it is almost needless to state—is now deserted. It is surrounded with fine large timber; there are several coffee-shops, a blacksmith's and wheelright's, and a neat little weather-board inn.

At this part, our German friends bade us farewell, to follow out their original plan of going to Forest Creek; they had persuaded four others to accompany them, so our number was reduced to fifteen, myself included. The scenery now became very beautiful, diversified with hill and dale, well wooded, with here and there a small creek, more agreeable to look at than to cross, as there were either no bridges or broken-down ones. The loveliness of the weather seemed to impart energy even to our horses; and we did not pitch our tents till we had travelled full sixteen miles. We were now close beside Mount Alexander, which is nearly covered with timber, chiefly white gum, wattle and stringy bark.

Friday, 17.—A lovely morning; we proceeded in excellent spirits, passing some beautiful scenery, though rather monotonous. During the first few miles, we went across many little creeks, in the neighbourhood of which were indications that the diggers had been at work. These symptoms we hailed with intense delight. Gregory told us the history of a hole in this neighbourhood, out of which five people cleared £13,000 worth of gold each in about a few hours. In lieu of sinking a shaft, they commenced in a gully (colonial for valley), and drove a hole on an inclined plane up the side of the hill or rise. However wet the season, they could never be inconvenienced, as the very inclination would naturally drain the hole. Such a precaution was not needed, as the whole party were perfectly satisfied with the success they had had without toiling for more. The country between here and the " Porcupine Inn" is exceedingly beautiful—not unlike many parts in the lowlands of Wales. About eight miles on the road we pass Barker's Creek, which runs through a beautiful vale.

We camped this evening about four or five
miles from Bendigo, and some miles from the
" Porcupine Inn," which we left behind us.
The " Porcupine" is a newly built inn on an
old spot, for I believe there was an inn in
existence there before the diggings were ever
heard or thought of. The accommodation
appears on rather a small seale. Near it
is a portion of the station of the Messrs.
Gibson, through which the public road runs;
some parts are fine, others wooded and
swampy.

Saturday, 18.—Fine day ; we now approached Bendigo. The timber here is very
large. Here we first beheld the majestic iron
bark, *Eucalypti*, the trunks of which are fluted
with the exquisite regularity of a Doric column ;
they are in truth the noblest ornaments of these
mighty forests. A few miles further, and the
diggings themselves burst upon our view.
Never shall I forget that scene, it well repaid
a journey even of sixteen thousand miles. The
trees had been all cut down; it looked like
a sandy plain, or one vast unbroken succession of
countless gravel pits—the earth was everywhere

turned up—men's heads in every direction were popping up and down from their holes. Well might an Australian writer, in speaking of Bendigo, term it " The Carthage of the Tyre of Forest Creek." The rattle of the cradle, as it swayed to and fro, the sounds of the pick and shovel, the busy hum of so many thousands, the innumerable tents, the stores with large flags hoisted above them, flags of every shape, colour, and nation, from the lion and unicorn of England to the Russian eagle, the strange yet picturesque costume of the diggers themselves, all contributed to render the scene novel in the extreme.

We hurried through this exciting locality as quickly as possible; and, after five miles travelling, reached the Eagle Hawk Gully, where we pitched our tents, supped, and retired to rest—though, for myself at least, not to sleep. The excitement of the day was sufficient cure for drowsiness. Before proceeding with an account of our doings at the Eagle Hawk, I will give a slight sketch of the character and peculiarities of the diggings themselves, which are of course not confined to one spot, but are the characteristics that usually exist in any auriferous regions, where the diggers are at work. I will leave myself, therefore, safely ensconced beneath a tent at the Eagle Hawk, and take a slight and rapid survey of the principal diggings in the neighbourhood from Saw-pit Gully to Sydney Flat.

CHAPTER VI.

THE DIGGINGS.

OF the history of the discovery of gold in Australia I believe few are ignorant; it is therefore necessary that my recapitulation of it should be as brief as possible. The first supposed discovery took place some sixty years ago, at Port Jackson. A convict made known to Governor Phillip the existence of an auriferous region near Sydney, and on the locality being examined, particles of real gold-dust were found. Every one was astonished, and several other spots were tried without success. Suspicion was now excited, and the affair underwent a thorough examination, which elicited the following facts. The convict, in the hope of obtaining his pardon as a reward, had filed a guinea and some brass buttons, which, judiciously mixed, made a tolerable pile of gold-dust, and this he carefully distributed over a small tract of sandy land. In lieu of the expected freedom, his ingenuity was rewarded with close confinement and other punishments. Thus ended the first idea of a gold-field in those colonies.

In 1841 the Rev. W. B. Clarke expressed his belief in the existence of gold in the valley of the Macquarie, and this opinion was greatly confirmed by the observations of European geologists on the Uralian Mountains. In 1849 an indisputable testimony was added to these opinions by a Mr. Smith, who was then engaged in some iron works, near Berrima, and who brought a splendid specimen of gold in quartz to the Colonial Secretary. Sir C. A. Fitzroy evinced little sympathy with the discovery, and in a despatch to Lord Grey upon the subject, expressed his opinion that " any investigation that the Government might institute with the view of ascertaining whether gold did in reality exist to any extent or value in

that part of the colony where it was supposed
from its geological formation that metal would
be found, would only tend to agitate the public
mind, &c."

Suddenly, in 1851, at the time that the
approaching opening of the Crystal Palace was
the principal subject of attention in England,
the colonies of Australia were in a state of far
greater excitement, as the news spread like
wild-fire, far and wide, that gold was really
there. To Edward Hammond Hargreaves be
given the honour of this discovery. This gentleman was an old Australian settler, just
returned from a trip to California, where he
had been struck by the similarity of the geological formation of the mountain ranges in his
adopted country to that of the Sacramento district. On his return, he immediately searched
for the precious metal; Ophir, the Turon, and
Bathurst well repaid his labour. Thus commenced the gold diggings of New South Wales.

The good people of Victoria were rather
jealous of the importance given by these events
to the other colony. Committees were formed,
and rewards were offered for the discovery of a
gold-field in Victoria. The announcement of

the Clunes Diggings in July, 1851, was the
result ; they were situated on a tributary of the
Loddon.   On September 8, those of Ballarat,
and on the 10th those of Mount Alexander
completely satisfied the most sceptical as to the
vast mineral wealth of the colony.   Bendigo
soon was heard of; and gully after gully succes-
sively attracted the attention of the public by
the display of their golden treasures.

The names given to these gullies open a
curious field of speculation.   Many have a sort
of digger's tradition respecting their first discovery.   The riches of Peg Leg Gully were
brought to light through the surfacing of three
men with wooden legs, who were unable to sink
a hole in the regular way.   Golden Gully was
discovered by a man who, whilst lounging on
the ground and idly pulling up the roots of
grass within his reach, found beneath one a
nest of golden nuggets.   Eagle Hawk derives
its name from the number of eagle-hawks seen
in the gully before the sounds of the pick and
shovel drove them away.   Murderer's Flat and
Choke'em Gully tell their own tale.   The Irish
clan together in Tipperary Gully.   A party of
South Australians gave the name of their chief

town to Adelaide Gully. The Iron Bark is so
called from the magnificent trees which abound
there.   Long, Piccaninny, and Dusty Gully
need no explanation.   The Jim Crow ranges
are appropriately so called, for it is only by
keeping up a sort of Jim Crow dancing move-
ment that one can travel about there ; it is the
roughest piece of country at the diggings.
White Horse Gully obtained its name from a
white horse whose hoofs, whilst the animal in a
rage was plunging here and there, flung up the
surface ground and disclosed the treasures
beneath.   In this gully was found the famous
" John Bull Nugget," lately exhibited in London.   The party to whom it belonged consisted
of three poor sailors ; the one who actually discovered it had only been a fortnight at the
diggings.   The   nugget   weighed   forty-five
pounds, and was only a few inches beneath the
surface.   It was sold for £5000 ; a good morning's work that !

Let us take a stroll round Forest Creek—
what a novel scene !—thousands of human
beings engaged in digging, wheeling, carrying,
and washing, intermingled with no little grumbling, scolding and swearing.   We approach

first the old Post-office Square; next our eye glances down Adelaide Gully, and over the Montgomery and White Hills, all pretty well dug up; now we pass the Private Escort Station, and Little Bendigo. At the junction of Forest, Barker, and Campbell Creeks we find the Commissioners' quarters—this is nearly five miles from our starting point. We must now return to Adelaide Gully, and keep alongside Adelaide Creek, till we come to a high range of rocks, which we cross, and then find ourselves near the head-waters of Fryer's Creek. Following that stream towards the Loddon, we pass the interesting neighbourhood of Golden Gully, Moonlight Flat, Windlass and Red Hill; this latter which covers about two acres of ground is so called from the colour of the soil, it was the first found, and is still considered as the richest auriferous spot near Mount Alexander. In the wet season, it was reckoned that on Moonlight Flat one man was daily buried alive from the earth falling into his hole. Proceeding north-east in the direction of Campbell's Creek, we again reach the Commissioners' tent.

The principal gullies about Bendigo are

Sailors's, Napoleon, Pennyweight, Peg Leg, Growler's, White Horse, Eagle Hawk, Californian, American, Derwent, Long, Picaninny, Iron Bark, Black Man's, Poor Man's, Dusty, Jim Crow, Spring, and Golden—also Sydney Flat, and Specimen Hill—Haverton Gully, and the Sheep-wash. Most of these places are well-ransacked and tunnelled, but thorough good wages may always be procured by tin dish washing in deserted holes, or surface washing.

It is not only the diggers, however, who make money at the Gold Fields. Carters, carpenters, storemen, wheelwrights, butchers, shoemakers, &c., usually in the long run make a fortune quicker than the diggers themselves, and certainly with less hard work or risk of life. They can always get from £1 to £2 a day without rations; whereas they may dig for weeks and get nothing. Living is not more expensive than in Melbourne : meat is generally from 4d. to 6d. a pound, flour about 1s. 6d a pound, (this is the most expensive article in housekeeping there,) butter must be dispensed with, as that is seldom less than 4s. a pound, and only successful diggers can indulge in such articles as cheese, pickles, ham, sardines, pickled salmon, or spirits, as all these things, though easily procured if you have gold to throw away, are expensive, the last-named article (diluted with water or something less innoxious) is only to be obtained for 30s. a bottle.

The stores, which are distinguished by a flag, are numerous and well stocked. A new style of lodging and boarding house is in great vogue. It is a tent fitted up with stringy bark couches, ranged down each side the tent, leaving a narrow passage up the middle. The lodgers are supplied with mutton, damper, and tea, three times a day, for the charge of 5s. a meal, and 5s. for the bed; this is by the week, a casual guest must pay double, and as 18 inches is on an average considered ample width to sleep in, a tent 24 feet long will bring in a good return to the owner.

The stores at the diggings are large tents, generally square or oblong, and everything required by a digger can be obtained for money, from sugar-candy to potted anchovies; from East India pickles to Bass's pale ale; from ankle jack boots to a pair of stays; from a baby's cap to a cradle; and every apparatus for mining, from a pick to a needle. But the

confusion—the din—the medley—what a scene
for a shop walker ! Here lies a pair of herrings
dripping into a bag of sugar, or a box of
raisins ; there a gay-looking bundle of ribbons
beneath two tumblers, and a half-finished bottle
of ale. Cheese and butter, bread and yellow
soap, pork and currants, saddles and frocks,
wide-awakes and blue serge shirts, green veils
and shovels, baby linen and tallow candles, are
all heaped indiscriminately together ; added to
which, there are children bawling, men swearing, store-keeper sulky, and last, not *least*,
women's tongues going nineteen to the dozen.

Most of the store-keepers are purchasers of
gold either for cash or in exchange for goods,
and many are the tricks from which unsuspect-
ing diggers suffer. One great and outrageous
trick is to weigh the parcels separately, or divide
the whole, on the excuse that the weight would
be too much for the scales ; and then, on adding
up the grains and pennyweights, the sellers often
lose at least half an ounce. On one occasion,
out of seven pounds weight, a party once lost
an ounce and three quarters in this manner.
There is also the old method of false beams—
one in favour of the purchaser—and here, unless

the seller weighs in both pans, he loses considerably. Another mode of cheating is to have glass pans resting on a piece of green baize; under this baize, and beneath the pan which holds the weights, is a wetted sponge, which causes that pan to adhere to the baize, and consequently it requires more gold to make it level; this, coupled with the false reckoning, is ruinous to the digger. In town, the Jews have a system of robbing a great deal from sellers before they purchase the gold-dust (for in these instances it must be *dust*): it is thrown into a zinc pan with slightly raised sides, which are well rubbed over with grease; and under the plea of a careful examination, the purchaser shakes and rubs the dust, and a considerable quantity adheres to the sides. A commoner practice still is for examiners of gold-dust to cultivate long finger-nails, and, in drawing the fingers about it, gather some up.

Sly grog selling is the bane of the diggings. Many—perhaps nine-tenths—of the diggers are honest industrious men, desirous of getting a little there as a stepping-stone to independence elsewhere; but the other tenth is composed of outcasts and transports—the refuse of Van

Diemen's Land—men of the most depraved and
abandoned characters, who have sought and
gained the lowest abyss of crime, and who
would a short time ago have expiated their
crimes on a scaffold.　They generally work or
rob for a space, and when well stocked with
gold, retire to Melbourne for a month or so,
living in drunkenness and debauchery. If, however, their holiday is spent at the diggings,
the sly grog-shop is the last scene of their
boisterous career.　Spirit selling is strictly pro-
hibited; and although Government will license a
respectable public-house on the *road*, it is
resolutely refused *on* the diggings.　The result
has been the opposite of that which it was
intended to produce.　There is more drinking
and rioting at the diggings than elsewhere,
the privacy and risk gives the obtaining it an
excitement which the diggers enjoy as much as
the spirit itself; and wherever grog is sold on
the sly, it will sooner or later be the scene of
a riot, or perhaps murder.　Intemperance is
succeeded by quarrelling and fighting, the
neighbouring tents report to the police, and the
offenders are lodged in the lock-up; whilst the
grog-tent, spirits, wine, &c., are seized and

taken to the Commissioners. Some of the
stores, however, manage to evade the law rather
cleverly—as spirits are not *sold*, " my friend"
pays a shilling more for his fig of tobacco, and
his wife an extra sixpence for her suet ; and they
smile at the store-man, who in return smiles
knowingly at them, and then glasses are brought
out, and a bottle produced, which sends forth
not a fragrant perfume on the sultry air.

It is no joke to get ill at the diggings ; doctors make you pay for it. Their fees are—
for a consultation, at their own tent, ten shillings ; for a visit out, from one to ten pounds,
according to time and distance. Many are
regular quacks, and these seem to flourish best.
The principal illnesses are weakness of sight,
from the hot winds and sandy soil, and dysen-
tery, which is often caused by the badly-cooked
food, bad water, and want of vegetables.

The interior of the canvas habitation of the
digger is desolate enough ; a box on a block of
wood forms a table, and this is the only furni-
ture ; many dispense with that. The bedding,
which is laid on the ground, serves to sit upon.
Diogenes in his tub would not have looked more
comfortless than any one else. Tin plates and

pannicans, the same as are used for camping
up, compose the breakfast, dinner, and tea service, which meals usually consist of the same
dishes—mutton, damper, and tea.

In some tents the soft influence of our sex is
pleasingly apparent : the tins are as bright as
silver, there are sheets as well as blankets on
the beds, and perhaps a clean counterpane, with
the addition of a dry sack or piece of carpet on
the ground ; whilst a pet cockatoo, chained
to a perch, makes noise enough to keep the
" missus " from feeling lonely when the good
man is at work. Sometimes a wife is at first
rather a nuisance; women get scared and fright-
ened, then cross, and commence a " blow up "
with their husbands ; but all their railing generally ends in their quietly settling down to this
rough and primitive style of living, if not without a murmur, at least to all appearance with
the determination to laugh and bear it. And
although rough in their manners, and not over
select in their address, the digger seldom wil-
fully injures a woman ; in fact, a regular Vandemonian will, in his way, play the gallant with
as great a zest as a fashionable about town—at
any rate, with more sincerity of heart.

Sunday is kept at the diggings in a very orderly manner; and among the actual diggers themselves, the day of rest is taken in a *verbatim* sense. It is not unusual to have an established clergyman holding forth near the Commissioners' tent, and almost within hearing will be a tub orator expounding the origin of evil, whilst a " mill " (a fight with fisticuffs) or a dog fight fills up the background.

But night at the diggings is the characteristic time: murder here—murder there—revolvers cracking — blunderbusses bombing — rifles going off—balls whistling—one man groaning with a broken leg—another shouting because he couldn't find the way to his hole, and a third equally vociferous because he has tumbled into one—this man swearing—another praying —a party of bacchanals chanting various ditties to different time and tune, or rather minus both. Here is one man grumbling because he has brought his wife with him, another ditto because he has left his behind, or sold her for an ounce of gold or a bottle of rum. Donnybrook Fair is not to be compared to an evening at Bendigo.

Success at the diggings is like drawing lottery tickets—the blanks far outnumber the prizes; still, with good health, strength, and above all perseverance, it is strange if a digger does not in the end reap a reward for his labour. Meanwhile, he must endure almost incredible hardships. In the rainy season, he must not murmur if compelled to work up to his knees in water, and sleep on the wet ground, without a fire, in the pouring rain, and perhaps no shelter above him more waterproof than a blanket or a gum tree; and this not for once only, but day after day, night after night. In the summer, he must work hard under a burning sun, tortured by the mosquito and the little stinging March flies, or feel his eyes smart and his throat grow dry and parched, as the hot winds, laden with dust, pass over him. How grateful now would be a draught from some cold sparkling streamlet; but, instead, with what sort of water must he quench his thirst? Much the same, gentle reader, as that which runs down the sides of a dirty road on a rainy day, and for this a shilling a bucket must be paid. Hardships such as these are often the daily routine of a digger's life; yet, strange to say, far from depressing the spirits or weakening the frame, they appear in most cases to give strength and energy to both. This is principally owing to the climate, which even in the wet season is mild and salubrious.

Perhaps nothing will speak better for the general order that prevails at the diggings, than the small amount of physical force maintained there by Government to keep some thousands of persons of all ages, classes, characters, religions and countries in good humour with the laws and with one another. The military force numbers 130, officers and men; the police about 300.

The Government escort is under the control of Mr. Wright, Chief Commissioner; it consists of about forty foot and sixty mounted police, with the usual complement of inspectors and sergeants; their uniform is blue with white facings, their head-quarters are by the Commissioners' tent, Forest Creek.

The private escort uniform is a plain blue frock coat and trowsers. It is under the superintendence of Mr. Wilkinson; the head-quarters are at Montgomery Hill, Forest Creek. Both these escorts charge one per cent for conveying gold.

F

For the Victoria diggings, there is a Chief Commissioner, one Acting Resident Commissioner; one Assistant Commissioner at Ballarat, one at Fryer's Creek, five at Forest Creek, and six at Bendigo.

Provision is made by Government for the support, at the mines, of two clergymen of each of the four State paid churches of England, Scotland, Rome, and Wesleyan, at a salary of £300 a year.

CHAPTER VII.

EAGLE HAWK GULLY.

BEFORE commencing an account of our operations at the Eagle Hawk, it will be necessary
to write a few words in description of our gold-
digging party there; their Christian names will
be sufficient distinction, and will leave their
incognito undisturbed.

This party, as I have said before, consisted
of five gentlemen, including my brother. Of
the latter I shall only say that he was young
and energetic, more accustomed to use his
brains than his fingers, yet with a robust frame,
and muscles well strengthened by the various
exercises of boating, cricketing, &c., with

which our embryo collegians attempt to prepare
themselves for keeping their " terms."

Frank —— (who, from being a married
man, was looked up to as the head of our rather
juvenile party) was of a quiet and sedate disposition, rather given to melancholy, for which
in truth he had cause. His marriage had
taken place without the sanction—or rather in
defiance of the wishes—of his parents, for his
wife was portionless, and in a station a few
grades, as they considered, below his own;
moreover, Frank himself was not of age. Private income, independent of his parents, he had
none. A situation as clerk in a merchant's
office was his only resource, and during three
years he had eked out his salary to support a
delicate wife—whose ill health was a never-
failing source of anxiety and expense—two
infants, and himself. An unexpected legacy
of £500 from a distant relative at last seemed
to open a brighter prospect before them; and
leaving his wife and children with their relatives,
he quitted England to seek in a distant land a
better home than all his exertions could procure
for them in their own country. I never felt
surprised or offended at his silent and pre-

occupied manner, accompanied at times by great depression of spirits, for it was an awful responsibility for one so young, brought up as he had been in the greatest luxury, as the eldest son of a wealthy merchant, to have not only himself but others nearest and dearest to maintain by his own exertions.

William ———, a tall, slight, and rather delicate-looking man, is the next of our party whom I shall mention. His youth had been passed at Christ's Hospital. This he quitted with the firm conviction (in which all his friends of course participated) that he had been greatly wronged by not having been elected a Grecian ; and a rich uncle, incited by the before-mentioned piece of injustice, took him under his care, and promised to settle him in the world as soon as a short apprenticeship to business had been gone through. A sudden illness put a stop to all these schemes. The physicians recommended change of air, a warmer climate, a trip to Australia. William had relatives residing in Melbourne, so the journey was quickly decided upon, a cabin taken ; and the invalid rapidly recovering beneath the exhilarating effects of the sea-breezes. How refreshing are they to the sick ! how caressingly does the soft sea-air fan the wan cheeks of those exhausted with a life passed amidst the brick walls and crowded, noisy streets of a city; and William, who at first would have laughed at so ridiculous a supposition, ere the four months' voyage was terminated, had gained strength and spirits sufficient to make him determine to undertake a trip to the diggings.

He was a merry light-hearted fellow, fonder of a joke than hard work, yet ever keeping a sharp eye to the "main chance," as the following anecdote will prove.

One day during our stay in Melbourne he came to me, and said, laughing :

"Well! I've got rid of one of the bad *habits* I had on board the ——."

"Which?" was my reply.

"That old frock-coat I used to wear in the cold weather whilst we rounded the Cape. A fellow down at Liardet's admired the cut, asked me to sell it. I charged him four guineas, and walked into town in my shirt-sleeves; soon colonized, eh ?"

Richard —— was a gay young fellow of twenty, the only son of a rich member of the

Stock Exchange.   In a fit of spleen, because
the parental regulations required him always to
be at home by midnight, he shipped himself off
to Australia, trusting that so energetic a step
"would bring the governor to his senses."   He
was music-mad, and appeared to know every
opera by heart, and wearied us out of all
patience with his everlasting humming of
" Ciascun lo dice" "Non piu mesta," &c.

   Octavius —— was the eighth son of a poor
professional man, who, after giving him a good
general education, sent him with a small capital
to try his fortune in the colonies.   For this he
was in every way well fitted, being possessed of
a strong constitution, good common sense, and
simple inexpensive habits; he was only nineteen, and the youngest of the male portion of
our party.

   The day after our arrival at the diggings,
being Sunday, we passed in making ourselves
comfortable, and devising our future plans.
We determined to move from our present
quarters, and pitch our tents a little higher up
the gully, near Montgomery's store.   This we
accomplished the first thing on Monday morning; and at about a hundred yards from us our

four shipmates also fixed themselves, which added both to our comfort and security.

A few words for their introduction.

One of them was a Scotchman, who wished to make enough capital at the mines to invest in a sheep-run; and as his countrymen are proverbially fortunate in the colonies, I think it possible he may some time hence be an Australian *millionaire*. Another of these was an architect, who was driven, as it were, to the diggings, because his profession, from the scarcity of labour, was at the time almost useless in Melbourne. The third was, or rather had been, a house-painter and decorator, who unfortunately possessed a tolerably fine voice, which led him gradually to abandon a good business to perform at concerts. Too late he found that he had dropped the substance for the shadow; emigration seemed his only resource; so leaving a wife and large. family behind, he brought his mortified vanity and ruined fortunes to begin the world anew with in Australia. He was the only one whose means prevented him from taking a share in our venture; but to avoid confusion, the Scotchman subscribed twice the usual sum, thus securing double profits. The fourth was a gentleman farmer, whose sole enemy, by his account, was Free Trade, and who held the names of Cobden and Bright in utter detestation.

As soon as the tents were pitched, all set to work to unpack the dray; and after taking out sufficient flour, sugar, tea, &c., for our own use, the remainder of the goods were taken to the nearest store, where they were sold at an average of five times their original cost: the most profitable portion of the cargo consisted of some gunpowder and percussion-caps. The day after, by great good fortune, we disposed of the dray and horses for £250, being only £40 less than we had paid for them. As the cost of keeping horses at the diggings is very great (sometimes two or three pounds a day per head), besides the constant risk of their being lost or stolen, we were well satisfied with the bargain; and never did nine young speculators, who five months previous had been utter strangers, accomplish their undertaking with greater satisfaction to themselves, or less disagreement one with another.

This business settled, the next was to pro-

cure licences, which was a walk of nearly five
miles to the Commissioners' tent, Bendigo, and
wasted the best part of Wednesday.

Meanwhile we were seriously debating about
again changing our quarters. We found it
almost impossible to sleep. Never before could
I have imagined that a woman's voice could
utter sounds sufficiently discordant to drive
repose far from us, yet so it was.

The gentlemen christened her " the amiable
female."

The tent of this "amiable" personage was
situated at right angles with ours and our shipmates, so that the annoyance was equally felt.
Whilst her husband was at work farther down
the gully, she kept a sort of sly grog-shop, and
passed the day in selling and drinking spirits,
swearing, and smoking a short tobacco-pipe at
the door of her tent. She was a most repulsive
looking object. A dirty gaudy-coloured dress
hung unfastened about her shoulders, coarse
black hair unbrushed, uncombed, dangled about
her face, over which her evil habits had spread
a genuine bacchanalian glow, whilst in a loud
masculine voice she uttered the most awful

words that ever disgraced the mouth of man—
ten thousand times more awful when proceeding
from a woman's lips.

But night was the dreadful time; then, if
her husband had been unlucky, or herself made
fewer profits during the day, it was misery to
be within earshot; so much so, that we decided to leave so uncomfortable a neighbourhood without loss of time, and carrying our
tents, &c., higher up the gully we finally
pitched them not far from the Portland
Stores.

This was done on Thursday, and the same
evening two different claims were marked out
ready to commence working the next day.
These claims were the usual size, eight feet
square.

Friday 24.—Early this morning our late
travelling companion, Joe, made his appearance
with a sack (full of bran, he said,) on his
shoulders. After a little confidential talk with
William, he left the sack in our tent, as he had
no other safe place to stow it away in till the
bran was sold. This gave rise to no suspicion,
and in the excitement of digging was quite
forgotten.

About noon I contrived to have a damper
and a large joint of baked mutton ready for the
" day labourers," as they styled themselves.
The mutton was baked in a large camp oven
suspended from three iron bars, which were
fixed in the ground in the form of a triangle,
about a yard apart, and were joined together at
the top, at which part the oven was hung over
a wood fire. This grand cooking machine was,
of course, outside the tent. Sometimes I have
seen a joint of meat catch fire in one of these
ovens, and it is difficult to extinguish it before
the fat has burnt itself away, when the meat
looks like a cinder.

Our butcher would not let us have less than
half a sheep at a time, for which we paid 8s.
I was not good housekeeper enough to know
how much it weighed, but the meat was very
good. Flour was then a shilling a pound, or
two hundred pounds weight for nine pounds in
money. Sugar was 1s. 6d., and tea 3s. 6d.
Fortunately we were well provided with these
three latter articles.

The hungry diggers did ample justice to the
dinner I had provided for them. They brought
home a tin-dish full of surface soil, which in

the course of the afternoon I attempted to
wash.

Tin-dish-washing is difficult to describe. It
requires a watchful eye and a skilful hand; it
is the most mysterious department of the gold-
digging business. The tin dish (which, of
course, is round) is generally about eighteen
inches across the top, and twelve across the
bottom, with sloping sides of three or four
inches deep. The one I used was rather
smaller. Into it I placed about half the "dirt"
—digger's technical term for earth, or soil—
that they had brought, filled the dish up with
water, and then with a thick stick commenced
making it into a batter; this was a most necessary commencement, as the soil was of a very
stiff clay. I then let this batter—I know no
name more appropriate for it—settle, and carefully poured off the water at the top. I now
added some clean water, and repeated the
operation of mixing it up; and after doing this
several times, the "dirt," of course, gradually
diminishing, I was overjoyed to see a few bright
specks, which I carefully picked out, and with
renewed energy continued this by no means
elegant work. Before the party returned to tea

I had washed out all the stuff, and procured from it nearly two pennyweights of gold-dust, worth about 6*s*. or 7*s*.

Tin-dish-washing is generally done beside a stream, and it is astonishing how large a quantity of "dirt" those who have the knack of doing it well and quickly can knock off in the course of the day. To do this, however, requires great manual dexterity, and much gold is lost by careless washing. A man once extracted ten pounds weight of the precious metal from a heap of soil which his mate had washed too hurriedly.

In the evening Joe made his re-appearance, carrying another sack on his shoulders, which contained a number of empty bottles, and now for the first time we became initiated into the *bran* mystery which had often puzzled us on the road—it seemed so strange a thing to carry up to the diggings. Joe laughed at our innocence, and denied having told us anything approaching a falsehood; a slight suppression of the truth was all he would plead guilty to. I verily believe William had put him up to this dodge, to make us smile when we should have felt annoyed. Being taxed with deceit, said he:

"I told you two-thirds truth; there wanted but two more letters to make it *brandy*," and with the greatest *sang-froid* he drew out a small keg of brandy from the first sack and half-filled the bottles with the spirit, after which he filled them all up to the neck with water. The bottles were then corked, and any or all of them politely offered to us at the rate of 30*s*. a piece. We declined purchasing, but he sold them all during the evening, for which we were rather glad, as, had they been discovered by the officials in our tent, a fine of £50 would have been the consequence of our foolish comrade's good-nature and joke-loving propensities.

We afterwards found that Master Joe had played the same trick with our shipmates and with the two doctors, who had bought a tent and settled themselves near our old place by Montgomery's store.

Saturday, 25.—The two holes were "bottomed" before noon with no paying result. It had been hard work, and they were rather low-spirited about it. The rest of the day they spent in washing some surface-soil, and altogether collected about an ounce and a half of gold-dust, counting the little I had washed out on the Friday. In the evening it was all dried by being placed in a spade over a quick fire. We had before determined to square accounts and divide the gold every Saturday night, but this small quantity was not worth the trouble, so it was laid by in the digger's usual treasury, a German match-box. These round boxes hold on an average eight ounces of gold.

These two unproductive holes had not been very deep. The top, or surface soil, for which a spade or shovel is used, was of clay. This was succeeded by a strata almost as hard as iron—technically called " burnt stuff,"—which robbed the pick of its points nearly as soon as the blacksmith had steeled them at a charge of 2s. 6d. a point. Luckily for their arms, this strata was but thin, and the yellow or blue clay which followed was comparatively easy work— here and there an awkward lump of quartz required the use of the pick. Suddenly they came to some glittering particles of yellow, which, with heartfelt delight they hailed as gold. *It was mica.* Many are at first deceived by it, but it is soon distinguished by its weight, as the mica will blow away with the slightest puff.

After a little useless digging among the clay, they reached the solid rock, and thus having fairly "bottomed" the holes to no purpose, they abandoned them.

Sunday, 26.—Although impossible at the diggings to keep this day with those outward observances which are customary in civilized life, we attempted to make as much difference as possible between the day of rest and that of work. Frank performed the office of chaplain, and read the morning service in the calm and serious manner which we expected from him.

I was rather amused to see the alacrity with which, when this slight service was over, they all prepared to assist me in the formation of a huge plum-pudding for the Sunday's dinner. Stoning plums and chopping suet seemed to afford them immense pleasure—I suppose it was a novelty; and, contrary to the fact implied in the old adage, "too many cooks spoil the broth," our pudding turned out A 1.

In the afternoon we strolled about, and paid a visit to our shipmates. I was certainly most agreeably surprised by the quiet and order that everywhere prevailed.

Monday, 27.—To-day our party commenced
"sinking" in a new spot at some little distance.
The first layer of black soil was removed, and
on some being washed in a tin dish, it was
found to contain a tolerable portion of gold, and
was pronounced to be worth transporting to the
tent to be regularly cradled. My first official
notice of this fact was from Richard, who
entered the tent humming "Suona la tromba,"
with a bucket full of this heavy soil in each
hand. He broke off in the middle of his song
to ask for some water to drink, and grumbled
most energetically at such dirty work. He
then gave me an account of the morning's
doings. After a thin layer of the black surface
soil, it appeared they had come to a strata of
thick yellow clay, in which gold was often very
abundant. This soil, from being so stiff, would
require "puddling," a work of which he did
not seem to relish the anticipation. Before the
day was over, a great number of buckets full of
both soils were brought up and deposited in
heaps near the tents. All, with the exception
of the "operatic" Richard, seemed in good
spirits, and were well satisfied with what had
been done in so short a time.

In the evening the other party of our ship-mates arrived, and were busy fixing their tent at a distance of about forty yards from us. Frank and the other four, though pretty tired with the day's labour, lent a helping hand, and the united efforts of the nine speedily accom-plished this business, after which an immense quantity of cold mutton, damper, and tea made a rapid disappearance, almost emptying my larder, which, by the bye, was an old tea-chest.

We asked our friends their motive for leaving the old spot, and they declared they could stand the " amiable female" no longer ; she grew worse and worse. " Her tongue was sich," observed the Scotchman, " as wad drive ony puir beastie wild." She had regularly quarrelled with the two doctors because they would not give her a written certificate, that the state of her health required the constant use of spirits. She offered them two guineas for it, which they indignantly refused, and she then declared her intention of injuring their practice as much as possible, which she had the power to do, as her tent was of an evening quite the centre of attrac-tion, and her influence proportionably great. Pity 'tis that such a woman should be able to mar or make the fortunes of her fellow-creatures.

Tuesday, 28.—The holes commenced yesterday were duly " bottomed," but no nice pocket-full of gold was the result; our shipmates, however, met with better success, having found three small nuggets weighing two to four ounces each at a depth of not quite five feet from the surface.

Wednesday, 29.—To-day was spent in puddling and cradling.

Puddling is on the same principle as tin-dish-washing, only on a much larger scale. Great wooden tubs are filled with the dirt and fresh water, and the former is chopped about in all directions with a spade, so as to set the metal free from the adhesive soil and pipe-clay. Sometimes I have seen energetic diggers tuck up their trowsers, off with their boots, step into the tub, and crush it about with their feet in the same manner as tradition affirms that the London bakers knead their bread. Every now and again the dirtied water is poured off gently, and with a fresh supply, which is furnished by a mate with a long-handled dipper from the stream or pool, you puddle away. The great thing is, not to be afraid of over-work, for the better the puddling is, so much the more easy and profitable is the cradling. After having been well beaten in the tubs, the "dirt" is put into the hopper of the cradle, which is then rocked gently, whilst another party keeps up a constant supply of fresh water. In the right hand of the cradler is held a thick stick, ready to break up any clods which may be in the hopper, but which a good puddler would not have sent there.

There was plenty of water near us, for a heavy rain during the night had filled several vacated holes, and as there were five pair of hands, we hoped, before evening, greatly to diminish our mud-heaps.

Now for an account of our proceedings.

Two large wooden tubs were firmly secured in the ground and four set to work puddling, whilst Frank busied himself in fixing the cradle. He drove two blocks into the ground; they were grooved for the rockers of the cradle to rest in, so as to let it rock with ease and regularity. The ground was lowered so as to give the cradle a slight slant, and thus enable the water to run off more quickly. If a cradle dips too much, a little gold may wash off with the light sand. The cradling machine, though simple in itself, is rather difficult to describe. In shape and size it resembles an infant's cradle, and over that portion of it where, if for a baby, a hood would be, is a perforated plate with wooden sides, a few inches high all round, forming a sort of box with the perforated plate for a bottom; this box is called the hopper. The dirt is here placed, and the constant supply of water, after well washing the stuff, runs out through a hole made at the foot of the cradle. The gold generally rests on a wooden shelf under the hopper, though sometimes a good deal will run down with the water and dirt into one of the compartments at the bottom, and to separate it from the sand or mud, tin-dish-washing is employed.

As soon as sufficient earth was ready, one began to rock, and another to fill the hopper with water. Richard continued puddling, William enacted Aquarius for him, whilst a fifth was fully occupied in conveying fresh dirt to the tubs, and taking the puddled stuff from them to the hopper of the cradle. Every now and then a change of hands was made, and thus passed the day. In the evening, the products were found to be one small nugget weighing a quarter of an ounce, and in gold-dust eight pennyweights, ten grains, being worth, at the digging price for gold, about thirty-five shillings. This was rather less than we had calculated upon, and Richard signified his intention of returning to Melbourne, " He could no longer put up with such ungentlemanly work in so very unintellectual a neighbourhood, with bad living into the bargain." These last words, which were pronounced *sotto voce*, gave us a slight clue to the real cause of his dislike to the diggings, though we did not thoroughly understand it till next morning. It originated in some bottles of mixed pickles which he had in vain wanted Frank, who this week was caterer for the party, to purchase at four shillings a bottle, which sum, as we were all on economical thoughts intent, Frank refused to expend on any unnecessary article of food. This we learnt next morning at breakfast, when Richard congratulated himself on that being the last meal he should make off tea, damper, and mutton, without the latter having something to render it eatable. The puddling and cradling work had, I fancy, given the finishing stroke to his disgust. Poor Dick! he met with little commiseration: we could not but remember the thousands in the old country who would have rejoiced at the simple fare he so much despised. William, in his laughing way, observed, " that he was too great a pickle himself, without buying fresh ones."

Richard left us on Thursday morning, and with him went one of the other party, the house-painter and decorator, who also found gold-digging not so pleasant as he had expected. We afterwards learnt that before reaching Kilmore they separated. Richard arrived safely in Melbourne, and entered a gold-broker's office at a salary of three pounds a week, which situation I believe he now fills ; and as " the governor," to use Richard's own expression, " has not yet come to his senses," he must greatly regret having allowed his temper to be the cause of his leaving the comforts of home. His companion, who parted with Richard at Kilmore, was robbed of what little gold he had, and otherwise maltreated, whilst passing through the Black Forest. On reaching Melbourne, he sold everything he possessed, and that not being sufficient, he borrowed enough to pay his passage back to England, where, doubtless, he will swell the number of those whose lack of success in the colonies, and vituperations against them, are only equalled by their unfitness ever to have gone there.

Thursday was past in puddling and cradling, with rather better results than on the first day, still it was not to our satisfaction, and on Friday two pits were sunk. One was shallow, and the bottom reached without a speck of gold making its appearance. The other was left over till the next morning. This was altogether very disheartening work, particularly as the expenses of living were not small. There were many, however, much worse off than ourselves, though here and there a lucky digger excited the envy of all around him. Many were the tricks resorted to in order to deceive new-comers. Holes were offered for sale, in which the few grains that were carefully placed in sight was all that the buyer gained by his purchase.

A scene of this description was enacted this Friday evening, at a little distance from us. The principal actors in it were two in number. One sat a little way from his hole with a heap

G

of soil by his side, and a large tin dish nearly
full of dirt in his hand.   As he swayed the dish
to and fro in the process of washing, an immense number of small nuggets displayed themselves, which fact in a loud tone he announced
to his " mate," at the same time swearing at
him for keeping at work so late in the evening.
This digger, who was shovelling up more dirt
from the hole, answered in the same elegant
language, calling him an " idle good-for-
nought."   Every now and then he threw a
small nugget to the tin-dish-washer, loudly
declaring, " he'd not leave off while them
bright bits were growing thick as taters underground."

"Then be d—d if I don't !" shouted the
other ; " and I'll sell the hole for two hundred
yeller boys down."

This created a great sensation among the
bystanders, who during the time had collected
round, and among whom was a party of three,
evidently " new chums."

" It shall go for a hundred and fifty !" again
shouted the washer, giving a glance in the
direction in which they stood.

" Going for a hundred, tin-dish as well !"

letting some of the water run off, and displaying the gold.

This decided the matter, and one of the three stepped forward and offered the required sum.

" Money down," said the seller; " these here fellers 'll witness it's all reg'lar."

The money was paid in notes, and the purchasers were about to commence possession by taking the tin-dish out of his hand.

" Wait till he's emptied. I promised yer the dish, but not the stuff in it," and turning out the " dirt" into a small tub the two worthies departed, carrying the tub away with them.

Not a grain of gold did the buyers find in the pit next morning.

Saturday, October 2.—This day found the four hard at work at an early hour, and words will not describe our delight when they hit upon a " pocket" full of the precious metal. The " pocket" was situated in a dark corner of the hole, and William was the one whose fossicking knife first brought its hidden beauties to light. Nugget after nugget did that dirty soil give up, by evening they had taken out five

pounds weight of gold.   Foolish Richard! we
all regretted his absence at this discovery.

As the next day was the Sabbath, thirty-six
hours of suspense must elapse before we could
know whether this was but a passing kindness
from the fickle goddess, or the herald of continued good fortune.

This night, for the first time, we were really
in dread of an attack, though we had kept our
success quite secret, not even mentioning it to
our shipmates; nor did we intend to do so until
Monday morning, when our first business would
be to mark out three more claims round the
lucky spot, and send our gold down to the
escort-office for security.   For the present we
were obliged to content ourselves with "plant-
ing" it—that is, burying it in the ground; and
not a footstep passed in our neighbourhood
without our imagining ourselves robbed of the
precious treasure, and as it was Saturday night
—the noisiest and most riotous at the diggings
—our panics were neither few nor far between.
So true it is that riches entail trouble and
anxiety on their possessor.

CHAPTER VIII.

AN ADVENTURE.

Sunday 3.—A FINE morning. After our usual service Frank, my brother, and myself, determined on an exploring expedition, and off we went, leaving the dinner in the charge of the others. We left the busy throng of the diggers far behind us, and wandered into spots where the sound of the pick and shovel, or the noise of human traffic, had never penetrated. The scene and the day were in unison; all was harmonious, majestic, and serene. Those mighty forests, hushed in a sombre and awful silence; those ranges of undulating hill and dale never yet trodden by the foot of man; the soft

still air, so still that it left every leaf unruffled,
flung an intensity of awe over our feelings, and
led us from the contemplation of nature to
worship nature's God.

We sat in silence for some while deeply im-
pressed by all around us, and, whilst still sitting
and gazing there, a change almost imperceptibly
came over the face of both earth and sky. The
forest swayed to and fro, a sighing moaning
sound was borne upon the wind, and a noise as
of the rush of waters, dark massive clouds
rolled over the sky till the bright blue heavens
were completely hidden, and then, ere we had
recovered from our first alarm and bewilderment, the storm in its unmitigated fury burst
upon us. The rain fell in torrents, and we
knew not where to turn.

Taking me between them, they succeeded in
reaching an immense shea-oak, under which
we hoped to find some shelter till the violence
of the rain had diminished; nor where we disappointed, though it was long before we could
venture to leave our place of refuge. At length,
however, we did so, and endeavoured to find
our way back to Eagle Hawk Gully. Hopeless
task ! The ground was so slippery, it was as

much as we could do to walk without falling;
the mud and dirt clung to our boots, and a
heavy rain beat against our faces and nearly
blinded us.

"It is clearing up to windward," observed
Frank; "another half-hour and the rain will
be all but over; let us return to our tree
again."

We did so. Frank was correct; in less
than the time he had specified a slight drizzling rain was all of the storm that remained.

With much less difficulty we again attempted
to return home, but before very long we made
the startling discovery that we had completely
lost our way, and to add to our misfortune the
small pocket-compass, which Frank had brought
with him, and which would have now so greatly
assisted us, was missing, most probably dropped
from his pocket during the skirmish to get
under shelter. We still wandered along till
stopped by the shades of evening, which came
upon us—there is little or no twilight in Australia.

We seated ourselves upon the trunk of a
fallen tree, wet, hungry, and, worst of all, ignorant of where we were. Shivering with cold,

and our wet garments hanging most uncomfortably around us, we endeavoured to console one another by reflecting that the next morning we could not fail to reach our tents. The rain had entirely ceased, and providentially for us the night was pitch dark—I say providentially, because after having remained for two hours in this wretched plight a small light in the distance became suddenly visible to us all, so distant, that but for the intensity of the darkness it might have passed unnoticed. "Thank God!" simultaneously burst from our lips.

"Let us hasten there," cried Frank, "a whole night like this may be your sister's death and would ruin the constitution of a giant."

To this we gladly acceded, and were greatly encouraged by perceiving that the light remained stationary. But it was a perilous undertaking. Luckily my brother had managed to get hold of a long stick with which he sounded the way, for either large stones or water-holes would have been awkward customers in the dark; wonderful to relate we escaped both, and when within hailing distance of the light, which we perceived came from a

torch held by some one, we shouted with all
our remaining strength, but without diminishing our exertions to reach it. Soon—with
feelings that only those who have encountered
similar dangers can understand—answering
voices fell upon our ears. Eagerly we pressed
forward, and in the excitement of the moment
we relinquished all hold of one another, and
attempted to wade through the mud singly.

"Stop! halt!" shouted more than one stentorian voice; but the warning came too late.
My feet slipped—a sharp pain succeeded by a
sudden chill—a feeling of suffocation—of my
head being ready to burst—and I remembered
no more.

When I recovered consciousness it was late
in the morning, for the bright sun shone upon
the ground through the crevices of a sail cloth
tent, and so different was all that met my eyes
to the dismal scene through which I had so
lately passed, and which yet haunted my memory,
that I felt that sweet feeling of relief which we
experience when, waking from some horrid
vision, we become convinced how unsubstantial
are its terrors, and are ready to smile at the pain
they excited.

G 3

That I was in a strange place became quickly
evident, and among the distant hum of voices
which ever and anon broke the silence not one
familiar tone could I recognize.   I endeavoured
to raise myself so as to hear more distinctly,
and then it was that an acute pain in the ankle
of the right foot, gave me pretty strong evidence
as to the reality of the last night's adventures.
I was forced to lie down again, but not before I
had espied a hand-bell which lay within reach
on a small barrel near my bed.   Determined as
far as possible to fathom the mystery, I rang a
loud peal with it, not doubting but what it
would bring my brother to me.   My surprise
and delight may be easier imagined than
described, when, as though in obedience to my
summons, I saw a small white hand push aside
the canvas at one corner of the tent, and one
of my own sex entered.

She was young and fair; her step was soft
and her voice most musically gentle.   Her eyes
were a deep blue, and a rich brown was the
colour of her hair, which she wore in very short
curls all round her head and parted on one side,
which almost gave her the appearance of a
pretty boy.

These little particulars I noticed afterwards; at that time I only felt that her gentle voice and kind friendliness of manner inexpressibly soothed me.

After having bathed my ankle, which I found to be badly sprained and cut, she related, as far as she was acquainted with them, the events of the previous evening. I learnt that these tents belonged to a party from England, of one of whom she was the wife, and the tent in which I lay was her apartment. They had not been long at the diggings, and preferred the spot where they were to the more frequented parts.

The storm of yesterday had passed over them without doing much damage, and as their tents were well painted over the tops, they managed to keep themselves tolerably dry; but later in the evening, owing to the softness of the ground, one of the side-posts partly gave way, which aroused them all, and torches were lit, and every one busied in trying to prop it up till morning. Whilst thus engaged, they heard our voices calling for help. They answered, at the same time getting ready some more torches before advancing to meet us, as there were

several pit-holes between us and them. Their
call for us to remain stationary came too late
to save me from slipping into one of their pits,
thereby spraining my ankle and otherwise hurt-
iny myself, besides being buried to my forehead
in mud and water. The pit was not quite five
feet deep, but, unfortunately for myself in this
instance, I belong to the pocket edition of the
feminine sex. They soon extricated me from
this perilous situation, and carried me to their
tents, where, by the assistance of my new friend,
I was divested of the mud that still clung to
me, and placed into bed.

Before morning the storm, which we all
thought had passed over, burst forth with re-
doubled fury; the flashes of lightning were
succeeded by loud peals of thunder, and the
rain came splashing down. Their tents were
situated on a slight rise, or they would have
run great risk of being washed away; every
hole was filled with water, and the shea-oak, of
whose friendly shelter we had availed ourselves
the evening before, was struck by lightning,
and shivered into a thousand pieces. After a
while the storm abated, and the warm sun and a
drying wind werequickly removing all traces of it.

Frank and my brother, after an early breakfast, had set out for Eagle Hawk Gully under the guidance of my fair friend's husband, who knew the road thither very well; it was only three miles distant. He was to bring back with him a change of clothing for me, as his wife had persuaded my brother to leave me in her charge until I had quite recovered from the effects of the accident, "which he more readily promised," she observed, "as we are not quite strangers, having met once before."

This awakened my curiosity, and I would not rest satisfied till fully acquainted with the how, when, and where. Subsequently she related to me some portion of the history of her life, which it will be no breach of confidence to repeat here.

Short as it is, however, it is deserving of another chapter.

CHAPTER IX.

HARRIETTE WALTERS.

HARRIETTE WALTERS had been a wife but twelve months, when the sudden failure of the house in which her husband was a junior partner involved them in irretrievable ruin, and threw them almost penniless upon the world. At this time the commercial advantages of Australia, the opening it afforded for all classes of men, and above all, its immense mineral wealth, were the subject of universal attention. Mr. Walters' friends advised him to emigrate, and the small sum saved from the wreck of their fortune served to defray the expenses of the journey.   Harriette, sorely against her

wishes, remained behind with an old maiden
aunt, until her husband could obtain a home
for her in the colonies.

The day of parting arrived; the ship which
bore him away disappeared from her sight,
and almost heart-broken she returned to the
humble residence of her sole remaining relative.

Ere she had recovered from the shock occasioned by her husband's departure, her aged
relation died from a sudden attack of illness,
and Harriette was left alone to struggle with
her poverty and her grief. The whole of her
aunt's income had been derived from an
annuity, which of course died with her; and
her personal property, when sold, realized not
much more than sufficient to pay a few debts
and the funeral expenses; so that when these
last sad duties were performed, Harriette found
herself with a few pounds in her pocket, home-
less, friendless, and alone.

Her thoughts turned to the distant land, her
husband's home, and every hope was centred in
the one intense desire to join him there. The
means were wanting, she had none from whom
she could solicit assistance, but her determina-

tion did not fail. She advertized for a situation as companion to an invalid, or nurse to young children, during the voyage to Port Philip, provided her passage-money was paid by her employer. This she soon obtained. The ship was a fast sailer, the winds were favourable, and by a strange chance she arrived in Melbourne three weeks before her husband. This time was a great trial to her. Alone and unprotected in that strange, rough city, without money, without friends, she felt truly wretched. It was not a place for a female to be without a protector, and she knew it, yet protector she had none ; even the family with whom she had come out, had gone many miles up the country. She possessed little money, lodgings and food were at an awful price, and employment for a female, except of a rough sort, was not easily procured.

In this dilemma she took the singular notion into her head of disguising her sex, and thereby avoiding much of the insult and annoyance to which an unprotected female would have been liable. Being of a slight figure, and taking the usual colonial costume—loose trowsers, a full, blue serge shirt, fastened round the waist by a leather belt, and a wide-awake—Harriette passed very well for what she assumed to be— a young lad just arrived from England. She immediately obtained a light situation near the wharf, where for about three weeks she worked hard enough at a salary of a pound a week, board, and permission to sleep in an old tumble-down shed beside the store.

At last the long looked-for vessel arrived. That must have been a moment of intense happiness which restored her to her husband's arms—for him not unmingled with surprise; he could not at first recognize her in her new garb. She would hear of no further separation, and when she learnt he had joined a party for the Bendigo diggings, she positively refused to remain in Melbourne, and she retained her boyish dress until their arrival at Bendigo. The party her husband belonged to had two tents, one of which they readily gave up to the married couple, as they were only too glad to have the company and in-door assistance of a sensible, active woman during their spell at the diggings. For the sake of economy, during the time that elapsed before they could commence their journey up, all of them lived in the

tents which they pitched on a small rise on the
south side of the Yarra.   Here it was that our
acquaintance first took place; doubtless, my
readers will, long ere this, have recognized in
the hospitable gentleman I encountered there,
my friend's husband, and, in the delicate-looking
youth who had so attracted my attention, the
fair Harriette herself.

But—*revenons à nos moutons.*

On the third day of my visit I was pronounced
convalescent, and that evening my brother and
William came to conduct me back to Eagle
Hawk Gully.   It was with no little regret that
I bade farewell to my new friend, and I must
confess that the pleasure of her society had
for the time made me quite careless as to
the quantity of gold our party might be taking
up during my absence. Whilst walking towards
our tents, I heard the full particulars of their
work, which I subjoin, so as to resume the
thread of my *digging* narrative in a proper
manner.

Monday.—Much upset by their anxiety occasioned by the non-appearance the previous

evening of Frank, my brother, and myself. The
two former did not reach home till nearly noon,
the roads were so heavy. After dinner all set
to work in better spirits; came to the end of
the gold—took out nearly four pounds weight.

Tuesday and Wednesday.—Digging various
holes in the vicinity of the lucky spot, but
without success. The other party did the same
with no better result.

Such were the tidings that I heard after my
three days' absence.

Thursday.—To-day was spent in prospecting
—that is, searching for a spot whose geological
formation gives some promise of the precious
metal. In the evening, William and Octavius
returned with the news that they had found a
place at some distance from the gully, which
they thought would prove "paying," as they
had washed some of the surface soil, which
yielded well. It was arranged that the party
should be divided into two, and take alternate
days to dig there.

Friday.—In pursuance of the foregoing plan,
William and Octavius set off, carrying a good
quantity of dinner and their tools along with
them. They worked hard enough during the

day, but only brought back three pennyweights of gold-dust with them. My brother and Frank gained a deal more by surface washing at home.

Saturday.—Changed hands. Frank and my brother to the new spot, digging. Octavius and William surface washing. The results were much the same as the day before.

Sunday, October 10.—We took advantage of the fine weather to pay a visit to Harriette and her party. We found them in excellent spirits, for at last they had hit upon a rich vein, which had for three days been yielding an average of four pounds weight a day, and was not yet exhausted. I say *at last*, for I have not before mentioned that they had never obtained more than an ounce of gold altogether, up to the day I left them. We were sincerely pleased with their good fortune. Harriette hoped that soon they might be able to leave this wild sort of life, and purchase a small farm, and once again have a home of their own. This could not be done near Melbourne, so they meant to go to South Australia, where any quantity of land may be bought. In *this* colony no smaller quantity than a square mile—640 acres—is sold by the

Government in one lot; consequently, those whose capital is unequal to purchase this, go to some other colony, and there invest the wealth they have acquired in Victoria.

As we had some idea of leaving Eagle Hawk Gully, I bade Harriette farewell. We never expected to meet again. It chanced otherwise; but I must not anticipate.

Monday and Tuesday were most unprofitably passed in digging holes; and on Tuesday night we determined to leave the Eagle Hawk, and try our fortune in some of the neighbouring gullies.

Wednesday was a bustling day. We sold our tent, tools, cradle, &c., as we knew plenty were always to be bought of those who, like ourselves, were changing their place. Had we known what we were about, we should never have burdened ourselves by bringing so many goods and chattels a hundred and twenty miles or more up the country; but " experience teaches." Having parted with all encumbrances, myself excepted, we started for the Iron Bark Gully. All the gold had been transmitted by the escort to Melbourne, and one fine nugget, weighing nearly five ounces, had been sent to

Richard. We could not resist the pleasure of presenting him with it, although by our rules not entitled to any of the proceeds.

The following are the rules by which our affairs were regulated. They were drawn up before leaving Melbourne, and signed by all. Though crude and imperfect, they were sufficient to preserve complete harmony and good fellowship between five young men of different character, taste, and education—a harmony and good fellowship which even Richard's withdrawal did not interrupt.

The rules were these:

1. No one party to be ruler; but every week by turn, one to buy, sell, take charge of gold, and transact all business matters.

2. The gold to be divided, and accounts settled every Saturday night.

3. Any one voluntarily leaving the party, to have one-third of his original share in the expense of purchasing tent and tools returned to him, but to have no further claim upon them or upon the gold that may be found after his withdrawal. Any one dismissed the party for misconduct, to forfeit all claim upon the joint property.

4. The party agree to stand by one another in all danger, difficulty, or illness.

5. Swearing, gambling, and drinking spirits to be strictly avoided.

6. Morning service to be read every Sunday morning.

7. All disputes or appeals from the foregoing rules to be settled by a majority.

CHAPTER X.

IRON BARK GULLY.

I HAVE said little in description of the Eagle Hawk, for all gullies or valleys at the diggings bear a strong external resemblance one to another. This one differed from others only in being much longer and wider; the sides, as is usually the case in the richest gullies, were not precipitous, but very gradual; a few mountains closed the background. The digging was in many places very shallow, and the soil was sometimes of a clayey description, sometimes very gravelly with slate bottom, sometimes gravelly with pipeclay bottom, sometimes quite sandy; in fact, the earth was of all sorts and depths.

At one time there were eight thousand diggers together in Eagle Hawk Gully. This was some months before we visited it. During the period of our stay at Bendigo there were not more than a thousand, and fewer still in the Iron Bark. The reasons for this apparent desertion were several.

The weather continued wet and uncertain, so that many who had gone down to Melbourne remained there, not yet considering the ground sufficiently recovered from the effects of the prolonged wet season, they had no desire to run the risk of being buried alive in their holes. Many had gone to the Adelaide diggings, of which further particulars hereafter, and many more had gone across the country to the Ovens, or, farther still, to the Sydney diggings themselves. According to digging parlance, "the Turon was looking up," and Bendigo, Mount Alexander, and Forest Creek were thinned accordingly. But perhaps the real cause of their desertion arose from the altered state of the diggings. Some time since one party netted £900 in three weeks; £100 a week was thought nothing wonderful. Four men found one day seventy-five pounds weight; another

H

party took from the foot of a tree gold to the value of £2000. A friend of mine once met a man whom he knew returning to Melbourne, walking in dusty rags and dirt behind a dray, yet carrying with him £1500 worth of gold. In Peg Leg Gully, fifty and even eighty pounds weight had been taken from holes only three or four feet deep. At Forest Creek a hole produced sixty pounds weight in one day, and forty more the day after. From one of the golden gullies a party took up the incredible quantity of one hundred and ninety-eight pounds weight in six weeks. These are but two or three instances out of the many that occurred to prove the richness of this truly auriferous spot. The consequence may be easily imagined; thousands flocked to Bendigo. The "lucky bits" were still as numerous, but being disseminated among a greater number of diggers, it followed that there were many more blanks than prizes, and the disappointed multitude were ready to be off to the first new discovery. Small gains were beneath their notice. I have often heard the miners say that they would rather spend their last farthing digging fifty holes, even if they found nothing in them, than "tamely" earn an ounce a day by washing the surface soil; on the same principle, I suppose, that a gambler would throw up a small but certain income to be earned by his own industry, for the uncertain profits of the cue or dice.

For ourselves, we had nothing to complain about. During the short space of time that we had been at Eagle Hawk Gully, we had done as well as one in fifty, and might therefore be classed among the lucky diggers; but "the more people have, the more they want;" and although the many pounds weight of the precious metal that our party had "taken up" gave, when divided, a good round sum a-piece, the avaricious creatures bore the want of success that followed more unphilosophically than they had done before the rich "pocketful" of gold had made its appearance. They would dig none but shallow holes, and a sort of gambling manner of setting to work replaced the active perseverance they had at first displayed.

Some days before we left, Eagle Hawk Gully had been condemned as a "worthless place," and a change decided on. The when and the where were fixed much in the following manner:

"I say, mates," observed William on the evening of the Sunday on which I had paid my last visit to Harriette, "I say, mates, nice pickings a man got last week in the Iron Bark— only twenty pounds weight out of one hole; that's all."

"Think it's true?" said Octavius, quietly.

"Of course; likely enough. I propose we pack up our traps, and honour this said gully with our presence forthwith."

"Let's inquire first," put in Frank; "it's foolish to change good quarters on such slight grounds."

"Good quarters! slight grounds!" cried William; "what next? what would you have? Good quarters! yes, as far as diggings concerned—whether you find anything for your digging is another matter. Slight grounds, indeed! twenty pounds weight in one day! Yes, we ought to inquire; you're right there, old boy, and the proper place to commence our inquiries is at the gully itself. Let's be off tomorrow."

"Wait two days longer," said Octavius "and I am agreeable."

And this, after a little chaffing between the impatient William and his more business-like comrades, was satisfactorily arranged.

Behold us then, on Wednesday the 13th, after having sold all our goods that were saleable, making our way to the Iron Bark Gully. William enacted the part of auctioneer, which he did in a manner most satisfactory to himself, and amusing to his audience; but the things sold very badly, so many were doing the same. The tents fetched only a few shillings each, and the tools, cradles, &c., *en masse*, were knocked down for half a sovereign.

The morning was rather cloudy, which made our pedestrian mode of travelling not so fatiguing as it might have been, had the sun in true colonial strength been shining upon us. This was very fortunately not the case, for we more than once mistook our way, and made a long walk out of a short one—quite a work of supererogation—for the roads were heavy and tiring enough without adding an extra quantity of them.

We passed in the close neighbourhood of

Sailor's, Californian, American, Long, and Piccaninny Gullies before reaching our destination.
Most of these gullies are considered ransacked,
but a very fair amount of gold-dust may be
obtained in either by the new-comer by tin-dish
fossicking in deserted holes.   These deserted
gullies, as they are called, contained in each no
trifling population, and looked full enough for
comfortable working.   What must they have
resembled the summer previous, when some
hundreds of people leaving a flat or gully was
but as a handful of sand from the sea-shore !

Before evening we arrived at the Iron Bark.
This gully takes its name from the splendid
trees with which it abounds ; and their immense
height, their fluted trunks and massive branches
gave them a most majestic appearance.   We
paused beneath one in a more secluded part,
and there determined to fix our quarters for the
night.   The heavy "swags" were flung upon
the ground, and the construction of something
resembling a tent gave them plenty to do ; the
tomahawks, which they carried in their belts,
were put into immediate requisition, and some
branches of the trees were soon formed into
rough tent-poles.   The tent, however, though

perhaps as good as could be expected, was
nothing very wonderful after all, being made
only of some of the blankets which our party
had brought in their swags.   Beneath it I
reposed very comfortably ; and, thanks to my
fatiguing walk, slept as soundly as I could
possibly have done beneath the roof of a palace.
The four gentlemen wrapped themselves in their
blankets, and laid down to rest upon the ground
beside the fire; their only shelter was the
foliage of the friendly tree which spread its
branches high above our heads.

Next morning William was for settling ourselves in the gully.   He wanted tents, tools,
&c., purchased, but by dint of much talking
and reasoning, we persuaded him first to look
well about, and judge from the success of others
whether we were likely to do any good by stopping there.   We soon heard the history of the
" twenty-pound weight " story.   As Frank and
Octavius had at once surmised, it originated in
a party who were desirous to sell their claims
and baggage before starting for Melbourne.   I
believe they succeeded—there are always plenty
of " new chums " to be caught and taken in—
and the report had caused a slight rush of

diggers, old and new, to the gully. Many of these diggers had again departed, others stayed to give the place a trial; we were not among the latter. The statements of those who were still working were anything but satisfactory, and we were all inclined to push on to Forest Creek.

Meanwhile, it is Thursday afternoon. All but Frank appear disposed for a siesta; he alone seems determined on a walk. I offer myself and am accepted as a companion, and off we go together to explore this new locality.

We proceeded up the gully. Deserted holes there were in numbers, many a great depth, and must have cost a vast amount of manual labour. In some places the diggers were hard at work, and the blows of the pick, the splash of water, and the rocking of the cradle made the diggings seem themselves again. There were several women about, who appeared to take as active an interest in the work as their " better halves." They may often be seen cradling with an infant in their arms. A man and a cart proceeded us up the gully. Every now and again he shouted out in a stentorian voice that made the welkin ring ; and the burden of his cry was this:

" 'Ere's happles, happles, Vandemonian happles, and them as dislikes the hiland needn't heat them."

The admirers of the fertile island must have been very numerous, for his customers soon made his pippins disappear.

We passed a butcher's shop, or rather tent, which formed a curious spectacle. The animals, cut into halves or quarters, were hung round; no small joints there—half a sheep or none; heads, feet, and skins were lying about for any one to have for the trouble of picking up, and a quantity of goods of all sorts and sizes, gridirons, saucepans, cradles, empty tea-chests, were lying scattered around in all directions ticketed " for sale." We quickly went on, for it was not a particularly pleasant sight, and at some distance perceived a quiet little nook rather out of the road, in which was one solitary tent. We hastened our steps, and advanced nearer, when we perceived that the tent was made of a large blanket suspended over a rope, which was tied from one tree to another. The blanket was fastened into the ground by large wooden pegs. Near to the opening of the tent, upon a piece of rock, sat a little girl of about ten years old. By her side was a quantity of the coarse green gauze of which the diggers' veils are made. She was working at this so industriously, and her little head was bent so fixedly over her fingers that she did not notice our approach. We stood for some minutes silently watching her, till Frank, wishing to see more of her countenance, clapped his hands noisily together for the purpose of rousing her.

She started, and looked up. What a volume of sorrow and of suffering did those pale features speak !

Suddenly a look of pleasure flashed over her countenance. She sprang from her seat, and advancing towards Frank, exclaimed:

"Maybe you'll be wanting a veil, Sir. I've plenty nice ones, stronger, better, and cheaper than you'll get at the store. Summer dust's coming, Sir. You'll want one, won't you? I havn't sold one this week," she added, almost imploringly, perceiving what she fancied a "no-customer" look in his face.

"I'll have one, little girl," he answered in a kindly tone, "and what price is it to be?"

"Eighteen pence, Sir, if you'd please be so good."

Frank put the money into her hand, but returned the veil. This action seemed not quite to satisfy her; either she did not comprehend what he meant, or it hurt her self-pride, for she said quickly :

" I havn't only green veils—p'raps you'd like some candles better—I makes them too."

" *You* make them?" said Frank, laughing as he glanced at the little hands that were still holding the veil for his acceptance. " *You* make them? Your mother makes the candles, you mean."

" I have no mother now," said she, with an expression of real melancholy in her countenance and voice. " I makes the candles and the veils, and the diggers they buys them of me, cos grandfather's ill, and got nobody to work for him but me."

" Where do you and your grandfather live?" I asked. " In there?" pointing to the blanket tent.

She nodded her head, adding in a lower tone :

" He's asleep now. He sleeps more than he did. He's killed hisself digging for the gold, and he never got none, and he says 'he'll dig till he dies.'"

"Dig till he dies." Fit motto of many a disappointed gold-seeker, the finale of many a broken up, desolated home, the last dying words of many a husband, far away from wife or kindred, with no loved ones near to soothe his departing moments—no better burial-place than the very hole, perchance, in which his last earthly labours were spent. These were some of the thoughts that rapidly chased one another in my mind as the sad words and still sadder tone fell upon my ear.

I was roused by hearing Frank's voice in inquiry as to how she made her candles, and she answered all our questions with a child-like *naïveté*, peculiarly her own. She told us how she boiled down the fat—how once it had caught fire and burnt her severely, and there was the scar still showing on her brown little arm—then how she poured the hot fat into the tin mould, first fastening in the wicks, then shut up the mould and left it to grow cold as quickly as it would; all this, and many other particulars which I have long since forgotten, she told us ; and little by little we learnt too her own history.

Father, mother, grandfather, and herself had all come to the diggings the summer before. Her father met with a severe accident in digging, and returned to Melbourne. He returned only to die, and his wife soon followed him to the grave. Having no other friend or relative in the colonies, the child had been left with her aged grandfather, who appeared as infatuated with the gold-fields as a more hale and younger man. His strength and health were rapidly failing, yet he still dug on. " We shall be rich, and Jessie a fine lady before I die," was ever his promise to her, and that at times when they were almost wanting food.

It was with no idle curiosity that we listened to her ; none could help feeling deeply interested in the energetic, unselfish, orphan girl. She was not beautiful, nor was she fair—she had none of those childish graces which usually attract so much attention to children of her age ; her eyes were heavy and bloodshot (with work, weeping, cold, and hunger) except when she spoke of her sick grandfather, and then they disclosed a world of tenderness ; her hair hung

matted round her head; her cheek was wan
and sallow; her dress was ill-made and thread-
bare; yet even thus, few that had once looked
at her but would wish to look again.   There
was an indescribable sweetness about the mouth;
the voice was low and musical; the well-shaped
head was firmly set upon her shoulders; a fine
open forehead surmounted those drooping eyes;
there was almost a dash of independence; a
"little woman" manner about her that made
one imperceptibly forget how young she was in
years.

A slight noise in the tent—a gentle moan.

" He's waked; I must go to him, and," in a
lower, almost a deprecating tone, " he doesn't
like to hear stranger folks about."

We cheerfully complied with the hint and
departed, Frank first putting some money into
her hand, and promising to call again for the
candles and veils she seemed quite anxious we
should take in return.

Our thoughts were as busy as our tongues
were silent, during the time that elapsed before
we reached home.   When we entered, we found
a discussion going on, and words were running
high.   My brother and Octavius were for going

somewhere to work, not idle about as they were doing now; William wanted to go for a " pleasure trip" to Forest Creek, and then return to Melbourne for a change. Frank listened to it all for some minutes, and then made a speech, the longest I ever heard from him, of which I will repeat portions, as it will explain our future movements.

" This morning, when going down the gully, I met the person whom we bought the dray-horses of in Melbourne. I asked him how he was doing, and he answered, ' badly enough; but a friend's just received accounts of some new diggings out Albury way, and there I mean to go.' He showed me also a letter he had received from a party in Melbourne, who were going there. From these accounts, gold is very plentiful at this spot, and I for one think we may as well try our fortune in this new place, as anywhere else. The route is partly along the Sydney road, which is good, but it is altogether a journey of two hundred miles. I would therefore propose (turning to my brother) that we proceed first to Melbourne, where you can leave your sister, and we can then start for the Ovens; and as provisions are at an exorbitant price

there, we might risk a little money in taking
up a dray-full of goods as before.   And as we
may never chance to be in this part of Victoria
again, I vote that we take William's 'pleasure
trip' to Forest Creek, stop there a few days,
and then to Melbourne."

This plan was adopted.

Friday morning.—Frank stole out early after
breakfast, for a visit to little Jessie.   I learnt
the full particulars afterwards, and therefore will
relate them as they occurred, as though myself
present.   He did not find her sitting outside
the tent as before, and hesitated whether to
remain or go away, when a low moaning inside
determined him to enter.   He pushed aside the
blanket, and saw her lying upon an old mat-
tress on the ground; beside her was a dark
object, which he could not at first distinguish
plainly.   It was her grandfather, and he was
dead.   The moaning came from the living
orphan, and piteous it was to hear her.   It
took Frank but a few minutes to ascertain all
this, and then he gently let down the blanket,
and hastened to the butcher's shop I have
already mentioned.   He learnt all that there
was to know: that she had no friends, no

relatives, and that nothing but her own labour, and the kindness of others, had kept them from starvation through the winter. Frank left a small sum in the butcher's hands, to have the old man buried, as best could be, in so wild and unnatural a place, and then returned to the mourning child. When he looked in, she was lying silent and senseless beside the corpse. A gentle breathing—a slight heaving of the chest, was all that distinguished the living from the dead. Carefully taking her in his arms, he carried her to our tent. As I saw him thus approaching, an idea of the truth flashed across me. Frank brought her inside, and laid her upon the ground—the only resting-place we had for her. She soon opened her eyes, the quick transition through the air had assisted in reviving her, and then I could tell that the whole sad truth returned fresh to her recollection. She sat up, resting her head upon her open hands, whilst her eyes were fixed sullenly, almost doggedly, upon the ground. Our attempts at consolation seemed useless. Frank and I glanced at one another. " Tell us how it happened," said he gently.

Jessie made no answer. She seemed like one who heard not.

"It must have been through some great carelessness—some neglect," pursued Frank, laying a strong emphasis on the last word.

This effectually roused her.

"I *never* left him—I *never* neglected him. When I waked in the morning I thought him asleep. I made my fire. I crept softly about to make his gruel for breakfast, and I took it him, and found him dead—dead," and she burst into a passion of tears.

Frank's pretended insinuation had done her good; and now that her grief found its natural vent, her mind became calmer, and exhausted with sorrow, she fell into a soothing slumber.

We had prepared to start before noon, but this incident delayed us a little. When Jessie awoke, she seemed to feel intuitively that Frank was her best friend, for she kept beside him during our hasty dinner, and retained his hand during the walk. There was a pleasant breeze, and we did not feel over fatigued when, after having walked about eight miles, we sat down beneath a most magnificent gum tree, more than a hundred feet high. Frank very wisely made Jessie bestir herself, and assist in our preparations. She collected dry sticks for a fire, went with him to a small creek near for a supply of water; and so well did he succeed, that for a while she nearly forgot her troubles, and could almost smile at some of William's gay sallies.

Next morning, very early, breakfast rapidly disappeared, and we were marching onwards. An empty cart, drawn by a stout horse, passed us.

Frank glanced at the pale little child beside him. " Where to ?" cried he.

" Forest Creek."

" Take us for what ?"

" A canary a-piece."

" Agreed." And we gladly sprung in. For the sake of the uninitiated, I must explain that, in digger's slang, a " canary " and half-a-sovereign are synonymous.

We passed the " Porcupine Inn." We halted at noon, dined, and about two hours after sighted the Commissioners' tent. In a few minutes the cart stopped.

"Can't take yer not no further. If the master seed yer, I'd cotch it for taking yer at all."

We paid him and alighted.

CHAPTER XI.

FOREST CREEK.

In my last chapter we were left standing not far from the Commissioners' tent, Forest Creek, at about three o'clock in the afternoon of Saturday, the 16th. An air of quiet prevailed, and made the scene unlike any other we had as yet viewed at the diggings. It was the middle of the month; here and there a stray applicant for a licence might make his appearance, but the body of the diggers had done so long before, and were disseminated over the creek digging, washing, or cradling, as the case might be, but here at least was quiet. To the right of the Licensing Commissioners' tent was

a large one appropriated to receiving the gold
to be forwarded to Melbourne by the Government escort.   There were a number of police
and pensioners about.

Not many months ago, the scarcity of these
at the diggings had prevented the better class
of diggers from carrying on their operations
with any degree of comfort, or feeling that their
lives and property were secure.   But this was
now altered; large bodies of police were placed
on duty, and wooden buildings erected in
various parts of the diggings for their accommodation.   Assistant Commissioners (who were
also magistrates) had been appointed, and large
bodies of pensioners enrolled as police, and
acting under their orders.   Roads were also
being made in all directions, thereby greatly
facilitating intercommunication.

But I must not forget that we are standing
looking about us without exactly knowing where
to turn.   Suddenly William started off like a
shot in pursuit of a man a little way from us.
We could not at first guess who it was, for in
the diggers' dress all men look like so many
brothers; but as we approached nearer we recognised our late captain, Gregory.

" Well, old fellow, and where did you spring
from ?" was Frank's salutation. " I thought
you were stuck fast in the Eagle Hawk."

" I may say the same," said Gregory, smiling.
" How got you here ?"

This was soon told, and our present dilemma
was not left unmentioned.

" A friend in need is a friend indeed," says
the proverb, and William echoed it, as Gregory
very complaisantly informed us that, having
just entered upon a store not far distant, he
would be delighted to give us a shelter for a
few nights. This we gladly accepted, and were
soon comfortably domiciled beneath a bark and
canvas tent adjoining his store. Here we
supped, after which Gregory left us, and returned with mattresses, blankets, &c., which he
placed on the ground, whilst he coolly ordered
the gentlemen to prepare to take their departure, he himself presently setting them the
example.

" I'm certain sure the young leddy's tired,"
said he ; " and that little lassie there (pointing
to Jessie) looks as pale and as wizened as an
old woman of seventy—the sooner they gets to
sleep the better."

We followed the kindly hint, and Jessie and myself were soon fast asleep in spite of the din close beside us. It was Saturday night, and the store was full; but the Babel like sounds disturbed us not, and we neither of us woke till morning.

It was Sunday. The day was fine, and we strolled here and there, wandering a good way from Gregory's store. As we returned, we passed near the scene of the monster meeting of 1851. The following account of it is so correct, that I cannot do better than transcribe it.

"The exceeding richness of the Mount Alexander diggings, and extraordinary success of many of the miners, led the Government to issue a proclamation, raising the licence from thirty shillings to three pounds. As soon as these intentions became known, a public meeting of all the miners was convened, and took place on the 15th of December, 1851. This resolve of the Governor and Executive Council was injudicious, since, in New South Wales, the Government proposed to reduce the fee to 15s.; and among the miners in Victoria, dissatisfaction was rife, on account of the apparent disregard by the Government of the wants and

wishes of the people engaged in the gold diggings,
and because of the absence of all police protection, while there appeared to be no effort made to
remedy this defect.   Indignation was, therefore,
unequivocally expressed at the several diggings'
meetings which were held, and at which it was
resolved to hold a monster meeting.   The
' Old Shepherd's Hut,' an out station of Dr.
Barker's, and very near the Commissioners'
tent, was the scene chosen for this display.
For miles around work ceased, cradles were
hushed, and the diggers, anxious to show their
determination, assembled in crowds, swarming
from every creek, gully, hill, and dale, even
from the distant Bendigo, twenty miles away.
They felt that if they tamely allowed the Government to charge £3 one month, the licensing
fee might be increased to £6 the next ; and by
such a system of oppression, the diggers' voca-
tion would be suspended.

" It has been computed that from fifteen to
twenty thousand persons were on the ground
during the time of the meeting.   Hundreds,
who came and heard, gave place to the coming
multitude, satisfied with having attended to

I

countenance the proceedings. The meeting
ultimately dispersed quietly, thereby disappoint-
ing the anticipations of those who expected,
perhaps even desired, a turbulent termination.
The majority determined to resist any attempt
to enforce this measure, and to pay *nothing ;*
but, happily, they were not reduced to this
extremity, since his Excellency wisely gave
notice that no change would be made in the
amount demanded for licence."

The trees up which the diggers had climbed
during the meeting are still pointed out.

The "Old Shepherd's Hut" was standing. It
seemed a most commodious little building compared to the insecure shelter of a digger's tent.
The sides of the hut were formed of slabs,
which were made mostly from the stringy bark,
—a tree that splits easily—the roof was composed of the bark from the same tree; the
chimney was of stones mortared together
with mud. This is the general style of
building for shepherds' huts in the bush. As
we passed it I could not but mentally contrast
the scene that took place there on the important
day of the monster meeting, to the deep tran-

quillity that must have reigned around the spot
for centuries before the discovery of gold drew
multitudes to the place.

The trees in this neighbourhood are mostly
stringy bark; almost all are peeled of their
covering, as many diggers, particularly those
who have their families with them, keep much
to one part, and think it, therefore, no waste
of time or labour to erect a hut, instead of
living in a comfortless tent.

On Monday morning we determined to
pursue our travels, and meant that day to pay
a flying visit to Fryer's Creek. It was a lovely
morning, and we set out in high spirits. A
heavy rain during the night had well laid the
dust. On our way we took a peep at several
flats and gullies, many of which looked very
picturesque, particularly one called Specimen
Gully, which was but thinly inhabited.

We had hardly reached Fryer's Creek
itself when we saw a vast concourse of people
gathered together. Frank and my brother
remained with me at a little distance, whilst
Octavius and William went to learn the occasion
of this commotion. It arose from an awful
accident which had just occurred.

Three brothers were working in a claim beside the stream, some way apart from the other diggers. The heavy rain during the night had raised the water, and the ground between the hole where they were working and the Creek, had given way imperceptibly *underneath.* One brother, who was early in the hole at work, fancied that the water at the bottom was gradually rising above his knees; he shouted to his comrades, but unfortunately they had gone, one, one way, one, another, in quest of something, and it was some minutes ere they returned.

Meanwhile the water in the hole was slowly but surely rising, and the slippery sides which were several feet high defied him to extricate himself. His cries for help became louder—he was heard, and his brothers and some neighbours hastened to his assistance. Ropes were procured after some further delay, and thrown to the unhappy man—but it was too late. None dared approach very near, for the ground was like a bog, and might at any moment give way beneath their feet; the water was nearly level with the top of the hole, and all hope of saving him was gone. The brothers had often been warned of the danger they were running.

Shuddering at the thoughts of this awful death we turned away, but no change of scene could dissipate it from our minds—the remembrance of it haunted me for many a night.

Jessie seemed pleased to see us on our return—we had left her behind with Gregory to his great delight—we abstained from mentioning before her the fearful accident we had all but witnessed.

That evening we wandered about Forest Creek. We had not gone far before a digger with a pistol in his hand shot by us; he was followed by an immense mob, hooting, yelling, and screaming, as only a mob at the diggings can. It was in full pursuit, and we turned aside only in time to prevent ourselves from being knocked down in the confusion.

"Stop him—stop him," was the cry. He was captured, and the cry changed to, "String him up—string him up—it's useless taking him to the police-office."

"What has he done?" asked my brother of a quiet by-stander.

"Shot a man in a quarrel at a grog-shop."

"String him up—string him up—confront him with the body," vociferated the mob.

At this moment the firmly-secured and well-guarded culprit passed by, to be confronted with the dead body of his adversary. No sooner did he come into his presence than the *ci-devant* corpse found his feet, "showed fight," and roared out, "Come on," with a most unghostlike vehemence. The fury of the mob cooled down; the people thought the man had been murdered, whereas the shot, fortunately for both, had glanced over the forehead without doing any serious injury. Taking advantage of this lull, the fugitive declared that the wounded man had been robbing him. This turned the tables, and, inspired by the hootings of the now indignant mob, the "dead man" took to his heels and disappeared.

The diggers in Pennyweight Flat, Nicholson's Gully, Lever Flat, Dirty Dick's Gully, Gibson's Flat, at the mouth of Dingley Dell, and in Dingley Dell itself, were tolerably contented with their gains, although in many instances,

the parties who were digging in the centre of
the gullies, or what is called " the slip," experienced considerable trouble in bailing the
water out of their holes.

Some of the names given to the spots about
Forest Creek are anything but euphonious.
Dingley Dell is, however, an exception, and
sounds quite musical compared to Dirty Dick's
Gully. The former name was given to the
place by a gentleman from Adelaide, and was
suggested by the perpetual tinkling of the
bullock's bells, it being a favourite camping-
place for bullock drivers, offering, as it did,
an excellent supply of both wood, water, and
food for their cattle. From whom the latter
inelegant name originated I cannot precisely tell
—but there are plenty of " dirty Dicks " all
over the diggings.

The current prices of this date at Forest
Creek were as follows : flour, £9 to £10 per
hundred-weight; sugar, 1s. 6d. a pound, very
scarce ; tea, 3s. ; rice, 1s. ; coffee, 3s. ; tobacco,
8s.; cheese, 3s. ; butter, 4s.; honey, 3s. 6d.;
candles, 1s. 6d ; currants, 1s. 6d., very scarce ;
raisins, 1s. 6d.; figs, 2s. 6d. ; salt, 1s. 6d.

Picks, spades, and tin dishes, 10s. each.　Gold 64s. per ounce.

Tuesday, 19.—Before breakfast we were busily employed in packing the " swags " when Octavius suddenly dropped the strap he held in his hand for that purpose, and darted into the store.　Thinking that we had omitted something which he went to fetch, we continued our work.　When everything was ready and the last strap in its place, we again thought of our absent comrade, making all sorts of surmises regarding his disappearance, when, just as Frank was going after him, in he walked, accompanied by a stranger whom he introduced as his uncle.　This surprised us, as we were ignorant of his having any relatives in the colonies.　He then explained that a younger brother of his father's had about eight years ago gone to South Australia, and that never having heard of him for some years they had mourned him as dead.　After many adventures he had taken a fancy to the diggings, and had just come from Melbourne with a dray full of goods.　He went to Gregory's store to dispose of them.　Octavius had heard them in conver-

sation together, and had mistaken his uncle's
for his father's voice. Hence the precipitation
of his exit. The uncle was a tall sunburnt
man, who looked well-inured to hardship and
fatigue. He stayed and took breakfast with us,
and then having satisfactorily arranged his
business with Gregory, and emptied his dray,
he obligingly offered to convey Jessie and myself
to Melbourne in it. Accordingly after dinner
we all started together.

Our new companion was a most agreeable
person, and his knowledge of the colonies was
extensive. With anecdotes of the bush, the
mines, and the town, he made the journey pass
most pleasantly. Before evening we reached
the Golden Point near Mount Alexander. This
term of " Golden " has been applied to a great
many spots where the deposits have been richer
than usual. There was a Golden Point at
Ballarat, and when the report of the Alexander
diggings drew the people from there, they
carried the name with them, and applied it to
this portion of the mount. To the left of the
Point, which was still full of labourers, was
the store of Mr. Black, with the Union-Jack
flying above it. It is a most noted store, and

at one time when certain delicacies were not to be had in Melbourne they were comparatively cheap here.

We passed by this busy spot and encamped at sunset at the foot of Mount Alexander. It was a lovely evening and our eyes were feasted by a most glorious sight. All the trees of the forest gradually faded away in the darkness, but beyond them, and through them were glimpses of the granite-like walls of the mount, brilliantly shining in and reflecting the last glowing rays of the setting sun. Some of the gorgeous scenes of fairy-land seemed before us—we could have imagined that we were approaching by night some illuminated, some enchanted castle.

That evening we sat late round our fire listening to the history which the uncle of Octavius related of some of his adventures in South Australia. The posts he had filled formed a curious medley of occupations, and I almost forget the routine in which they followed one another, but I will endeavour to relate his story as much as possible in his own words.

" When I started from England, after having paid passage-money, &c., I found myself with about £200 ready money in my purse—it was all I had to expect, and I determined to be very careful of it ; but by a young man of five-and-twenty these resolutions, like lady's promises, are made to be broken. When I landed in Adelaide with my money in my pocket —minus a few pounds I had lost at whist and cribbage on board ship—I made my way to the best inn, where I stayed some days, and ran up rather a longish bill. Then I wanted to see the country, which I found impossible without a horse, so bought one, and rode about to the various stations, where I was generally hospitably received, and thus passed a few months very pleasantly, only my purse was running low. I sold the horse, then my watch, and spent the money. When that was gone, I thought of the letters of introduction I possessed. The first that came to hand was directed to a Wesleyan minister. I called there, looking as sanctimonious as I could. He heard my story, advised me to go to chapel regularly, 'And for your temporal wants,' said he, 'the Lord will provide.' I thanked him, and bowed myself off.

"My first act was to burn my packet of

introductory letters, my next was to engage myself to a stock-holder at 15*s*. a week and my rations. He was going up to his station at once, and I accompanied him. We travelled for about two hundred miles through a most beautiful country before we reached his home. His house was, in my ideas, a comical-looking affair—made of split logs of wood, with a bark roof, and a barrel stuck on the top of the roof at one end by way of a chimney-pot. His wife, a pale sickly little woman, seemed pleased to see us, for she had been much alarmed by the natives, who were rather numerous about the neighbourhood. There was only a young lad, and an old shepherd and his wife upon the station, besides herself. Before I had been there six weeks she died, and her new-born little baby died too ; there was not a doctor for miles, and the shepherd's wife was worse than useless. I believe this often happens in the bush—it's not a place for woman-folks.

" I was here eighteen months—it was a wild sort of life, and just suited my fancy; but when I found I had some money to receive, I thought a spree in town would be a nice change, so off I marched. My spree lasted as long as my

money, and then I went as barman to a public-
house at Clare, some way up the country —
here I got better wages and better board, and
stopped about half-a-year. Then I turned
brewer's drayman, and delivered casks of good
Australian ale about Adelaide for 30s. a week.
The brewer failed, and I joined in a speculation
with an apple dealer to cart a lot up to the
Kapunda copper mines. That paid well. I
stopped up there as overseer over four-and-
twenty bullock-drays. Well, winter came, and
I had little to do, though I drew my 30s. a
week regularly enough, when the directors
wanted a contract for putting the small copper-
dust into bags, and sewing them up. I offered
to do the job at 2d. a bag, and could get
through a hundred and fifty a day. How much
is that? Oh! 12s. 6d. a-piece. I forgot to
tell you I'd a mate at the work. That was
good earnings in those days; and me and my
mate, who was quite a lad, were making a
pretty penny, when some others offered to do
them a halfpenny a bag cheaper. I did the
same, and we kept it to ourselves for about four
weeks longer, when a penny a bag was offered.
There was competition for you! This roused

my bile—I threw it up altogether—and off to Adelaide again. Soon spent all my cash, and went into a ship-chandler's office till they failed; then was clerk to a butcher, and lost my situation for throwing a quarter of his own mutton at him in a rage; and then I again turned brewer's man. Whilst there I heard of the diggings—left the brewer and his casks to look after themselves, and off on foot to Ballarat.

"Here I found the holes averaging some thirty feet—which was a style of hard work I didn't quite admire; so hearing of the greater facility of the Alexander diggings, I went through Bully Rook Forest, and tried my luck in the Jim Crow Ranges. This paid well; and I bought a dray, and bring up goods to the stores, which I find easier work, and twice as profitable as digging. There's my story; and little I thought when I went into Gregory's store to-day, that I should find my curly-pated nephew ready to hear it."

Next day we travelled on, and halted near Saw-pit Gully; it was early in the afternoon, and we took a walk about this most interesting locality. The earth was torn up everywhere—

a few lucky hits had sufficed to re-collect a
good many diggers there, and they were working vigorously. At dusk the labour ceased—
the men returned to their tents, and for the last
time our ears were assailed by the diggers' usual
serenade. Imagine some hundreds of revolvers
almost instantaneously fired—the sound rever-
berating through the mighty forests, and echoed
far and near—again and again till the last faint
echo died away in the distance. Then a hundred blazing fires burst upon the sight—around
them gathered the rough miners themselves—
their sun-burnt, hair-covered faces illumined
by the ruddy glare. Wild songs, and still
wilder bursts of laughter are heard ; gradually
the flames sink and disappear, and an oppressive
stillness follows (sleep rarely refuses to visit the
diggers' lowly couch), broken only by some
midnight carouser, as he vainly endeavours to
find his tent. No fear of a "peeler" taking him
off to a police-station, or of being brought
before a magistrate next morning, and "fined
five shillings for being drunk."

Early on Tuesday morning I gave a parting
look to the diggings — our dray went slowly

onwards—a slight turn in the road, and the last tent has vanished from my sight. "Never," thought I, " shall I look on such a scene again !"

CHAPTER XII.

RETURN TO MELBOURNE.

BEFORE the evening of Wednesday the 20th, we passed through Kyneton, and found our-selves in the little village of Carlshrue, where we passed the night.   Here is a police-station, a blacksmith's, a few stores and some cottages, in one of which we obtained a comfortable supper and beds.   A lovely view greeted us at sunrise.   Behind us were still towering the lofty ranges of Mount Alexander, before us was Mount Macedon and the Black Forest. This mountain, which forms one of what is called the Macedon range, is to be seen many miles distant, and on a clear, sunny day, the

purple sides of Mount Macedon, which stands
aloof as it were, from the range itself, are
distinctly visible from the flag-staff at Melbourne.

We had intended to have stopped for the
night in Kyneton, but the charges there were
so enormous that we preferred pushing on
and taking our chance as to the accommodation
Carlshrue could afford, nor did we repent the
so doing.

The following are the Kyneton prices. A
meal or bed—both bad—4s.; a night's stabling,
£1 10s. per horse; hay at the rate of 9d. a
pound; this is the most exorbitant charge of
all.

Hay was somewhere about £20 a ton in
Melbourne. The carriage of it to Kyneton,
now that the fine weather was setting in, would
not exceed £8 a ton at the outside, which would
come to £28. The purchaser, by selling it at
Kyneton at the rate of 9d. a pound, or £75
per ton, cleared a profit of £47—*not quite* 200
per cent. If *this* is not fortune-making, I
should like to know what is. It beats the
diggings hollow.

Next morning we looked our last at " sweet

Carlshrue," and having crossed the Five Mile-
Creek, camped for our mid-day meal beside the
Black Forest. Here a slight discussion arose,
as to whether it would be more advisable to
proceed on our journey and camp in the Black
Forest that night, or whether we should remain
where we were outside, and re-commence our
journey in good time the next morning so as
to get through this most uncomfortable portion
of our travels in one day. Frank and Octavius
were for the latter plan, as the best and
safest, but the rest (thinking that, having once
travelled through it without encountering anything resembling a bushranger, they might
safely do so again) protested against wasting
time, and were for entering those dark shades
without further delay. The uncle of Octavius
whom, in future, for the sake of convenience,
I shall call Mr. L——, was also of this mind,
and as he was in some sort our leader during
the journey, his advice decided the matter.
Danger to him was only a necessary excitement.
He was naturally fearless, and his merry laugh
and gay joke at the expense of the bushranger-
fearing party gradually dissipated the unac-

countable presentiment of danger which I for one had in no small degree experienced.

On we went, up hill and down dale, sometimes coming to a more open piece of ground, but more generally threading our way amid a very maze of trees, with trunks all black as the ground itself, whilst the dingy foliage and the few rays of sunshine that lit up those dark, deep glades served only to heighten the gloominess around.

After walking for about six miles—I preferred that mode of getting along to the joltings of the dray—we all felt disposed to rest ourselves. We selected a spot where the trees were less thickly clustered, and taking the horses out of the dray, tethered them by strong ropes to some trees near. The dray itself was turned up, and a blanket thrown over the up-raised shafts formed a most complete and cosy little tent.

A fire was next kindled, and a kettle full of water (with the tea in it!) was placed on to boil, some home-made bread, brought from Carlshrue, was placed upon the ground, and some chops were toasted on the ends of sticks,

which are usually the impromptu toasting-forks
of the bush. The old tin plates and pannicans,
not quite so bright as once upon a time, but
showing, despite sundry bruises and scratches,
that they had seen better days, were placed
upon the tea-table, which of course was the
ground. Two or three knives and forks were
on general service, and wandered about from
hand to hand as occasion required. Altogether
it was a merry, sociable party, and I think I
enjoyed that supper better than any I ever
tasted before or since.

" Chacun à sa gout," many a one will say.

The pleasantest moments must come to
an end, and so did these. After having sat
up later than usual, Jessie and I retired to
our gipsy tent, leaving our guardian diggers
smoking round the fire. They meant to
keep watches during the night to prevent a
surprise.

Friday.—We were comfortably seated at our
breakfast, discussing a hundred subjects besides
the food before us, when a shrill " coo-ey" burst
through the air; " coo-ey"—" coo-ey" again
and again, till the very trees seemed to echo
back the sound. We started to our feet, and,

as if wondering what would come next, looked
blankly at each other, and again the " coo-ey,"
more energetic still, rang in our ears. This is
the call of the bush, it requires some little skill
and practice, and when given well can be heard
a great way off. In such a place as the Black
Forest it could only proceed from some one
who had lost their way, or be a signal of distress
from some party in absolute danger. We again
looked from one to the other—it bewildered us;
and again the cry, only more plaintive than
before, came to us. Simultaneously they seized
their pistols, and started in the direction whence
the sounds proceeded. They were all too true
Englishmen to hear a fellow-creature in peril
and not hasten to their succour.

Jessie and myself could not remain behind
alone—it was impossible; we followed at a
little distance, just keeping our comrades in
sight. At last they came to a halt, not knowing where to turn, and we joined them. Frank
gave a " coo-ey," and in about the space of a
minute the words "help, help,—come, come,"
in scarcely audible sounds, answered to the call.
We penetrated about thirty yards farther, and
a few low groans directed us to a spot more

obscure, if possible, than the rest.   There, firmly bound to two trees close together, were two men.   A thick cord was passed round and round their bodies, arms, and legs, so as to leave no limb at liberty.   They seemed faint and exhausted at having called so long for help.

It was the work of a moment for our party to fling down their pistols, take out knives and tomahawks, and commence the work of releasing them from their bonds.   But the cords were knotted and thick, and there seemed no little labour in accomplishing it.   They were also retarded by the small quantity of light, for, as I said before, it was a dark and secluded spot. At length one man was released, and so faint and exhausted was he, from the effects of whatever ill-usage he had suffered, that, being a tall, powerfully made man, it required the united strength of both Frank and Mr. L——— to prevent his falling to the ground.

Jessie and myself were standing a little apart in the shade ; we seemed as if spell-bound by the incident, and incapable of rendering any assistance.

The second was soon set at liberty, and

no sooner did he feel his hands and feet free from the cords than he gave a loud, shrill " coo-ey."

A shriek burst from Jessie's lips as, immediately the cry was uttered, and before any one could recover from the bewilderment it occasioned, four well-armed men sprang upon our startled party.

Taken thus at disadvantage, unarmed, their very knives flung down in their eagerness to untwist the cords, they were soon overpowered. The wretch who had been reclining in Frank's arms quickly found his feet, and, ere Frank could recover from his surprise, one heavy blow flung him to the ground; whilst the other twined his powerful arms round Mr. L——, and, after a short but sharp struggle, in which he was assisted by a fellow-villain, succeeded in mastering him.

It was a fearful sight, and I can hardly describe my feelings as I witnessed it. My brain seemed on fire, the trees appeared to reel around me, when a cold touch acted as a sudden restorative, and almost forced a scream from my lips. It was Jessie's hand, cold as marble, touching mine. We spoke together in a low whisper, and both seemed inspired by the same thoughts, the same hope.

" I saw a little hill as we came here," said Jessie; " let's try and find it and look out for help."

I instinctively followed her, and stealthily creeping along, we gained a small rise of ground which commanded a more extended view than most places in the Black Forest, and, but for the thickness of the trees, we could have seen our own camping-place, and the part where the ambuscade had been laid.   From the sounds of the voices, we could tell that the ruffians were leading their prisoners to the spot where we had passed the night, and the most fearful oaths and imprecations could ever and anon be heard.   Well might our hearts beat with apprehension, for it was known that, when disappointed in obtaining the gold they expected, they vented their rage in torturing their unfortunate victims.

Meanwhile Jessie seemed listening intently. The time she had spent in the bush and at the diggings had wonderfully refined her sense of hearing.   Suddenly she gave a shrill " coo-ey." The moment after a shot was fired in the

K

direction of our late camp. Jessie turned even paler, but recovering herself, " coo-ey " after " coo-ey " made the echoes ring. I joined my feeble efforts to hers; but she was evidently well used to use this peculiar call. On a fine still day, this cry will reach for full three miles, and we counted upon this fact for obtaining some assistance.

" Help is coming," said Jessie, in a low voice, and once more with increasing strength she gave the call.

Footsteps approached nearer and nearer. I looked up, almost expecting to see those villainous countenances again.

" Women in danger!" shouted a manly voice, and several stalwart figures bounded to our side.

" Follow, follow!" cried Jessie, rushing forwards. I scarcely remember everything that occurred, for I was dizzy with excess of pleasure. There was a short scuffle, shots were fired at retreating bushrangers, and we saw our friends safe and free.

The whole matter was then related to our preservers—for such they were—and I then learnt that when the bushrangers had marched

off our party to the camping-place, they proceeded to overhaul their pockets, and then
bound them securely to some trees, whilst one
stood ready with a pistol to shoot the first that
should call for help, and the others looked over
the plunder. This was little enough, for our
travelling money, which was notes, was kept
—strange treasury—in the lining of the body
of my dress, and here too were the gold receipts
from the Escort Office. Every night I took
out about sufficient to defray the day's expenses,
and this was generally given into Frank's
hands.

Enraged and disappointed, the villains used
most frightful language, accompanied by threats
of violence; and the one on guard, irritated
beyond his powers of endurance, fired the pistol
in the direction of William's head. At this
moment Jessie's first "coo-ey" was heard:
this startled him, and the shot, from the aim
of the pistol being disarranged, left him un-
hurt.

"It's that d—d child," muttered one, with a
few additional oaths; "we'll wring her neck
when we've secured the plunder."

One of the ruffians now attempted more per-

suasive measures, and addressing Mr. L——,
whom I suppose he considered the leader, ex-
pended his powers of persuasion much in the
following manner.

"You sees, mate, we risks our lives to get
your gold, and have it we will. Some you've
got somewhere or another, for you havn't none
on you got no paper from the Escort—you
planted it last night, eh? Jist show us where,
and you shan't be touched at all, nor that little
wretch yonder, what keeps screeching so; but
if you don't—" and here his natural ferocity
mastered him, and he wound up with a volley
of curses, in the midst of which our rescuers
rushed upon them.

When we came to talk the whole matter over
calmly and quietly, no doubt was left upon our
minds, as to the premeditation of the whole
affair. But for the watch kept, the attack
would most probably have been made during
the night.

Our timely friends were a party of successful
diggers returning from work. They too had
passed the night in the Black Forest—providentially not very far from us. They accepted
our thanks in an off-hand sort of way, only

replying—which was certainly true—"that we would have done the same for them." It was in endeavouring to assist assumed sufferers that our party fell into the ambuscade laid for them.

They waited whilst we got the dray and horses ready, and we all journeyed on together, till the Black Forest was far behind us. We saw no more of the bushrangers, and encamped that night a few miles beyond the "Bush Inn." At this inn we parted with our gallant friends. They were of the jovial sort, and having plenty of gold, were determined on a spree. We never met them again.

On Saturday we travelled as far as the "Deep Creek Inn." Some distance before reaching that place, we passed two rival coffee-shops on the road. We stopped at the first, to know if they had any uncooked or cold meat to sell, for our provisions were running low.

"Havn't none," said the woman, shaking her head. Then looking hard at William, and judging from his good-humoured face, that he was a likely one to do what she wanted, she said to him: "Now, Sir, I'm agoing to ax a favour of you, and that is to go a little farther

down the road, to the other coffee-tent, and buy
for me as much meat as they'll let you have.
They's got plenty, and I've none; and they
knows I'll lose custom by it, so you'll not get
it if they twigs *(Anglicè* guesses) you comes
from me.   You understand, Sir," and she put
a sovereign into his hand to pay for it.

Laughing at the comicality of the request,
and the thoroughly colonial coolness of making
it, William set off, and presently returned with
nearly half a sheep hanging over his shoulders,
and a large joint in one hand.

"Bless me, what luck!" exclaimed the delighted woman, and loud and profuse were her
thanks.   She wanted to cook us a good dinner
off the meat gratis; but this we steadily
refused, and purchasing enough for the present,
we put our drays again into motion, and a little
while after kindled a fire, and were our own
cooks as usual.   That night we camped beside
the Deep Creek, about a mile from the " Deep
Creek Inn."   The route we were now taking
was different to the one we had travelled going
up—it was much more direct.

We remained all Sunday beside the creek, and
the day passed quietly and pleasantly.

On Monday the 25th we were again in motion.   We passed the well-known inn of Tulip Wright's.   How great a change those few weeks had made !   Winter had given place to summer, for Australia knows no spring. We walked along the beautiful road to Flemington, gave a look at the flagstaff and cemetery, turned into Great Bourke Street, halted at the Post-office, found several letters, and finally stopped opposite the " Duke of York Hotel," where we dined.

I shall leave myself most comfortably located here, whilst I devote a chapter or two to other diggings.

CHAPTER XIII.

BALLARAT.

BALLARAT is situated about forty-five miles from Geelong, and seventy-five nearly west of Melbourne. This was the first discovered gold-field of any extent in Victoria, and was made known on the 8th of September, 1851. The rush from Geelong was immense. Shops, stores, trades, all and everything was deserted; and the press very truly declared that " Geelong was mad—stark, staring gold-mad." During the month of September five hundred and thirty-two licences were taken out; in the month following the number increased to two thousand two hundred and sixty one!

The usual road to Ballarat is by the Adelaide overland route on the Gambier Road; but the most preferable is per Geelong. The former route leads over the Keilor Plains, and through Bacchus Marsh, crossing the Werribee River in two places. Mount Buninyong then appears in sight of the well-pleased traveller, and Ballarat is soon reached.

The route *viâ* Geelong is much quicker, as part of the way is generally performed by steam at the rate of £1 a-piece. Those who wish to save their money go to Geelong by land. After leaving Flemington, and passing the Benevolent Asylum, the Deep Creek is crossed by means of a punt, and you then come to a dreary waste of land, called Iett's Flat. Beyond is a steep rise and a barren plain, hardly fit to graze sheep upon, and at about twenty miles from Melbourne you come to the first halting house. Some narrow but rapid creeks must be got over, and for seven miles further you wander along over a dreary sheep-run till stopped by the Broken River, which derives its name partly from the nature of its rocky bed, and partly from the native name which has a similar sound.

This creek is the most steep, rapid, and dangerous on the road, having no bridge and no properly defined crossing-place or ford, except the natural rocks about. The bottom is of red sand-stone and rocks of the same description abut from the sides of the creek, and appear to abound in the neighbourhood; and all along the plains here and there are large fragments of sand and lime-stone rocks. Two hundred yards from the creek is a neat inn after the English style, with a large sitting-room, a tap, a bar, and a coffee-room. The bed-rooms are so arranged as to separate nobs from snobs—an arrangement rather inconsistent in a democratic colony. The inn also affords good stabling and high charges. Up to this distance on our road there is a scarcity of wood and springs of water.

We now pass two or three huts, and for twenty miles see nothing to please the eye, for it is a dead, flat sheep-walk. About seven miles on the Melbourne side of Geelong, the country assumes a more cheering appearance—homesteads, gardens, and farms spring up—the roads improve, and the timber is plentiful and large, consisting of shea-oaks, wattle, stringy

bark, and peppermints. Many of the houses
are of a good size, and chiefly built of stone,
some are of wood, and very few of brick.

Geelong, which is divided into north and
south, is bounded by the Barwin, a river navi-
gable from the bay to the town, and might be
extended further ; beautiful valleys well wooded
lie beyond. Between the two townships a park
has been reserved, though not yet enclosed ;
the timber in it, which is large—consisting
principally of white gum and stringy bark—is
not allowed to be cut or injured. There are
several good inns, a court-house, police-station,
and corporation offices. There is also a neat
church in the early pointed style, with a parsonage and schools in the Elizabethan ; all are
of dark lime-stone, having a very gloomy appearance, the stones being unworked, except
near the windows ; the porches alone are
slightly ornamented. The road and pavement
are good in the chief streets ; there is a large
square with a conduit, which is supplied by an
engine from the Barwin. The shops are large
and well furnished, a great many houses are
three stories high, most are two, and very few
one. The best part of the town is about one

hundred feet above the river. A large timber bridge over the Ballarat road was washed down last winter. The town is governed by a mayor and corporation. There is a city and mounted police force, and a neat police-court. A large and good race-course is situated about three miles from the town.

As regards scenery, Geelong is far superior to Melbourne, the streets are better, and so is the society of the place; none of the ruffian gangs and drunken mobs as seen in Victoria's chief city. There are various chapels, schools, markets, banks, and a small gaol. The harbour is sheltered, but not safe for strangers, as the shoals are numerous. Geelong is surrounded by little townships. Irish Town, Little Scotland, and Little London are the principal; and to show how completely the diggings drained both towns and villages of their male inhabitants, I need only mention that six days after the discovery of Ballarat, there was only one man left in Little Scotland, and he was a cripple, compelled *nolens volens* to remain behind.

The road from Geelong to Ballarat is well marked out, so often has it been trodden; and

there are some good inns on the way-side for
the comfort of travellers. On horseback you
can go from the town to the diggings in six
or eight hours.

Ballarat is a barren place, the ground is
interspersed with rocky fragments, the creek
is small, and good water is rather scarce. In
summer it almost amounts to a drought, and
what there is then is generally brackish or stag-
natic. It is necessary never to drink stagnant
water, or that found in holes, without boiling,
unless there are frogs in it, then the water is
good; but the diggers usually boil the water,
and add a drop of brandy, if they can get it. In
passing through the plains you are sure of
finding water near the surface (or by seeking a
few inches) wherever the tea tree grows.

The chief object at the Ballarat diggings is
the Commissioners' tent, which includes the
Post-office. There are good police quarters
now. The old lock-up was rather of the primitive order, being the stump of an old tree, to
which the prisoners were attached by sundry
chains, the handcuff being round one wrist and
through a link of the chain. I believe there is
now a tent for their accommodation. There

are several doctors about, who, as usual, drive a rare trade.

It is almost impossible to describe accurately the geological features of the gold diggings at Ballarat. Some of the surface-washing is good, and sometimes it is only requisite to sink a few feet, perhaps only a few inches, before finding the ochre-coloured earth (impregnated with mica and mixed with quartzy fragments), which, when washed, pays exceedingly well. But more frequently a deep shaft has to be sunk.

Of course the depth of the shafts varies considerably; some are sixty or even eighty, and some are only ten feet deep. Sometimes after heavy rains, when the surface soil has been washed from the sides of the hills, the mica layer is similarly washed down to the valleys and lies on the original surface-soil. This constitutes the true washing stuff of the diggings. Often when a man has—to use a digger's phrase—"bottomed his hole," (that is, cut through the rocky strata, and arrived at the gold layer), he will find stray indications, but nothing remunerative, and perchance the very next hole may be the most profitable on the diggings. Whether there is any geological rule to be guided by has yet to be proved, at present no old digger will ever sink below the mica soil, or leave his hole until he arrives at it, even if he sinks to forty feet. So, therefore, it may be taken as a general rule, wherever the diggings may be, either in Victoria, New South Wales, or South Australia, that gold in "working" quantities lies only where there is found quartz or mica.

Ballarat has had the honour of producing the largest masses of gold yet discovered. These masses were all excavated from one part of the diggings, known as Canadian Gully, and were taken out of a bed of quartz, at the depths of from fifty to sixty-five feet below the surface. The deep indentures of the nuggets were filled with the quartz. The largest of these masses weighed one hundred and thirty-four pounds, of which it was calculated that fully one hundred and twenty-six pounds consisted of solid gold !

About seven miles to the north of Ballarat, some new diggings called the Eureka have been discovered, where it appears that although there are no immense prizes, there are few blanks, and every one doing well !

In describing the road from Melbourne to

Geelong, I have made mention of the Broken River. A few weeks after my arrival in the colonies this river was the scene of a sad tragedy.

I give the tale, much in the same words as it was given to me, because it was one out of many somewhat similar, and may serve to show the state of morality in Melbourne.

The names of the parties are, of course, entirely fictitious.

———

Prettiest among the pretty girls that stood upon the deck as the anchor of the Government immigrant ship ' Downshire ' fell into Hobson's Bay, in August, 1851, was Mary H——, the heroine of my story. No regret mingled with the satisfaction that beamed from her large dark eyes, as their gaze fell on the shores of her new country, for her orphan brother, the only relative she had left in their own dear Emerald Isle, was even then preparing to follow her. Nor could she feel sad and lonely whilst the rich Irish brogue, from a subdued but manly and well-loved voice, fell softly on her ear, and the gentle pressure of her hand continually reminded her that she was not alone.

Ship-board is a rare place for match-making, and, somehow or another, Henry Stephens had contrived to steal away the heart of the 'Downshire' belle. Prudence, however, compelled our young people to postpone their marriage, and whilst the good housewife qualities of the one readily procured her a situation in a highly respectable family in Melbourne, Henry obtained an appointment in the police force of the same town.

Their united savings soon mounted up, and in a few months the banns were published, and Christmas-Day fixed on for the wedding. Mary, at her lover's express desire, quitted her mistress's family to reside with a widow, a distant relative of his own, from whose house she was to be married. Delightful to the young people was this short period of leisure and uninterrupted intercourse, for the gold mania was now beginning to tell upon the excited imaginations of all, and Henry had already thrown up his situation; and it was settled their wedding trip should be to the golden gullies round Mount Buninyong.

And now let me hasten over this portion of
my narrative. It is sad to dwell upon the
history of human frailty, or to relate the oft-told
tale of passion and villainy triumphant over
virtue. A few days before Christmas, when
the marriage ceremony was to be performed,
they unfortunately spent one evening together
alone, and he left her—ruined. Repentance
followed sin, and the intervening time was
passed by Mary in a state of the greatest
mental anguish. With what trembling eagerness did she now look forward to the day which
should make her his lawful wife.

It arrived. Mary and the friends of both
stood beside the altar, whilst he, who should
have been there to redeem his pledge and save
his victim from open ruin and disgrace, was far
away on the road to Ballarat.

To describe her agony would be impossible.
Day after day, week after week, and no tidings
from him came; conscience too acutely accounting to her for his faithlessness. Then the
horrible truth forced itself upon her, that its consequences would soon too plainly declare her sin
before the world; that upon her innocent offspring
would fall a portion of its mother's shame.

Thus six months stole sorrowfully away, and as yet none had even conjectured the deep cause she had for misery. Her brother's non-arrival was also an unceasing source of anxiety, and almost daily might she have been seen at the Melbourne Post-office, each time to return more disappointed than before. At length the oft-repeated inquiry was answered in the affirmative, and eagerly she tore open the long-anticipated letter. It told her of an unexpected sum of money that had come into his hands—to them a small fortune—which had detained him in Ireland. This was read and almost immediately forgotten, as she learnt that he was arrived in Melbourne, and that only a few streets now separated them.

She raised her face, flushed and radiant with joyful excitement—her eyes fell upon him who had so cruelly injured her. The scream that burst from her lips brought him involuntarily to her side. What will not a woman forgive where once her heart has been touched—in the double joy of the moment the past was almost forgotten—together they re-read the welcome letter, and again he wooed her for his bride. She consented, and he himself led her to her

brother, confessed their mutual fault, and
second preparations for an immediate marriage
were hurriedly made.

Once more at the altar of St. Peter's stood
the bridal party, and again at the appointed
hour Stephens was far gone on his second expedition to the diggings, after having increased
(if that was possible) his previous villainy, by
borrowing a large portion of the money before
mentioned from his intended brother-in-law.
It was pretty evident that the prospect of doing
this had influenced him in his apparently hon-
ourable desire to atone to the poor girl, who,
completely prostrated by this second blow, was
laid on the bed of sickness.

For some weeks she continued thus, and her
own sufferings were increased by the sight of
her brother's fury, as, on her partial recovery,
he quitted her in search of her seducer.

During his absence Mary became a mother,
and the little one that nestled in her bosom,
made her half forgetful of her sorrows, and at
times ready to embrace the delusive hope that
some slight happiness in life was in store for
her. But her bitter cup was not yet drained.
Day by day, hour by hour, her little one pined

away, until one dreary night she held within her arms only its tiny corpse.

Not one sound of grief—not an outward sign to show how deeply the heart was touched—escaped her. The busy neighbours left her for awhile, glad though amazed at her wondrous calmness; when they returned to finish their preparations for committing the child to its last resting-place, the mother and her infant had disappeared.

Carrying the lifeless burden closely pressed against her bosom, as though the pelting rain and chilling air could harm it now, Mary rapidly left the town where she had experienced so much misery, on—on—towards Geelong, the route her seducer and his pursuer had taken—on—across Iett's Flat, until at length, weak and exhausted, she sank down on the barren plains beyond.

Next morning the early dawn found her still plodding her weary way—her only refreshment being a dry crust and some water obtained at an halting-house on the road; and many a passer-by, attracted by the wildness of her eyes, her eager manner, and disordered dress, cast after her a curious wondering look.   But she heeded them not—on—on she pursued her course towards the Broken River.

Here she paused.   The heavy winter rains had swollen the waters, which swept along, dashing over the irregular pieces of rock that formed the only means of crossing over.   But danger was as nothing to her now—the first few steps were taken—the rapid stream was rushing wildly round her—a sensation of giddiness and exhaustion made her limbs tremble—her footing slipped on the wet and slimy stone—in another moment the ruthless waters carried her away.

The morrow came, and the sun shone brightly upon the still swollen and rapid river.   Two men stood beside it, both too annoyed at this impediment to their return to Melbourne to be in the slightest degree aware of their proximity to one another.   A bonnet caught by a projecting fragment of rock simultaneously attracted their attention : both moved towards the spot, and thus brought into closer contact they recognized each other.   Deadly foes though they were, not a word passed between them, and silently they dragged the body of the unhappy girl to land. In her cold and tightened grasp still lay the child.

As they stood gazing on those injured ones, within one breast remorse and shame, in the other, hatred and revenge, were raging violently.

Each step on the road to Ballarat had increased her brother's desire for vengeance, and still further was this heightened on discovering that Stephens had already left the diggings to return to town. This disappointment maddened him; his whole energy was flung into tracing his foe, and in this he had succeeded so closely, that unknown to either, both had slept beneath the same roof at the inn beside the Broken River.

The voices of some of the loungers there, who were coming down to the Creek to see what mischief had been done during the night, aroused him. He glanced upon his enemy, who pale and trembling, stood gazing on the wreck that he had made. Revenge at last was in his hands—not a moment was to be lost—with the yell of a maniac he sprang upon the powerless and conscious-stricken man—seized him in his arms—rushed to the river—and ere any could interpose, both had found a grave where but a few minutes before the bodies of Mary and her infant had reposed.

CHAPTER XIV.

NEW SOUTH WALES.

ABOUT seventy years ago a small colony of convicts first made the forests ring with the blows of the axe, and a few tents were erected where Sydney now stands. The tents, and they who dwelt beneath them, have long since disappeared, and instead we have one of the finest cities that our colonial empire ever produced.

The streets in Sydney are, as in Melbourne, built at right angles with one another; they are macadamized, well lighted with gas, and perambulated by a number of policemen during

L

the night. Some of the shops almost rival those of London, and the public buildings are good and numerous.   There is a custom-house, a treasury, police-office, college, benevolent asylum, banks, barracks, hospitals, libraries, churches, chapels, a synagogue, museum, club-house, theatre, and many splendid hotels, of which the largest is, I think, the "Royal Hotel," in George Street, built ąt the cost of £30,000.

Hyde Park is close at hand, with un-num-bered public walks, and a botanical garden, the favourite resort of all classes.

In the neighbourhood of Sydney are some good oyster-beds, and many are the pic-nics got up for the purpose of visiting them.   The oysters cling to the rocks, and great numbers are easily obtained.

The distance from Sydney to Melbourne, by the overland road, is about six hundred miles ; but the steamers, which are constantly plying, afford a more comfortable mode of transit.

The gold diggings of New South Wales are

so well known as to require but a cursory notice.
The first official notification of the fact of gold
having been discovered bears date, May 22,
1851, and is contained in a despatch from the
Governor to Earl Grey. In it he announced
the existence of a gold field to the westward of
Bathurst, about one hundred and fifty miles
from Sydney. At the same time, he added his
supposition that the gold sent for inspection was
Californian.

Mr. Stutchbury, the geological surveyor,
quickly undeceived his Excellency. He wrote
from Summer Hill Creek reporting that four
hundred persons were hard at work, and that
the gold existed not only in the creek but
beyond it. The following postscript was added
to his letter: " Excuse this being written in
pencil, as there is no ink in this city of Ophir."
And this appropriate name has ever since been
retained.

The natural consequences of this discovery
was the flocking of hundreds of the inhabitants
of Sydney to Bathurst. Sober people began to
be alarmed at this complete *bouleversement* of

business and tranquillity.   For the sake of
order the Governor attempted to put a stop to
the increasing desertion of the capital by pro-
claiming that the gold-fields were the preroga-
tive of the Crown, and threatening gold-diggers
with prosecution.   It was all in vain.   The
glitterings of the precious metal were more
attractive than the threats of the Governor
were otherwise.   The people laughed good-
humoured at the proclamation, and only
flocked in greater numbers to the auriferous
spot.

Government now took a wiser course, and
finding it impossible to stem the torrent, determined to turn the eagerness of the multitude
to some account.   A licence-fee of 30s., or
half an ounce of gold, per month was imposed,
which, with few exceptions, has always been
cheerfully paid.

The Turon diggings soon followed those of
Bathurst.   This river flows into the Macquarie
after a course of a hundred miles.   Along the
entire length auriferous discoveries are constantly
being made, and it bids fair to last for many

years to come. The gold is not found, as
many erroneously suppose, so much among the
sand as by digging in the soil. It also exists
in paying quantities on the shores and in the
beds of the Macquarie, the Abercrombie, and
Belubula rivers. Major's Creek, too, is a
favourite locality, and was first made known by
a prospecting woman.

According to Mr. Stutchbury's report, he
found gold *almost wherever he tried for it*,
and whilst traversing the Macquarie from
Walgumballa to the Turon, he found it at
every place he tried. Surely Midas must, once
upon a time, have taken a pleasure-trip to
Australia!

The delirium of the Sydney gold-fever
reached its height when it became publicly
known that a piece of one hundred and six
pounds weight had been disembowelled from
the earth at one time. This immense quantity
was the discovery of a native, who, being excited by the universal theme of conversation,
provided himself with a tomahawk, and ex-

plored the country adjacent to his employer's land. He was attracted by a glittering yellow substance on the surface of a block of quartz. With his tomahawk he broke off a piece, which he carried home to his master, Dr. Kerr, of Wallawa. Not being able to move the mass conveniently, Dr. Kerr broke it into small fragments. The place where it was found is at the commencement of an undulating table-land, very fertile, and near to a never-failing supply of water in the Murroo Creek. It is distant about fifty miles from Bathurst, thirty from Wellington, and twenty from the nearest point of the Macquarie river.

Dr. Kerr presented the native and his brother with two flocks of sheep, two saddle-horses, a quantity of rations, a team of bullocks, and some land.

About twenty yards from the spot where this mass was found, a piece of gold called the " Brennan Nugget " was soon after discovered. It weighed three hundred and thirty-six ounces, and was sold in Sydney for more than £1100.

But it would be useless to enter into fuller particulars of the diggings of New South Wales. Panoramas, newspapers, and serials have made them familiar to all.

CHAPTER XV.

SOUTH AUSTRALIA.

ADELAIDE, the capital of South Australia, was
the last formed of the three sister colonies.   In
1834 an act of colonization was obtained; and
land, both in town and country, sold rapidly.
The colonists, however, were most unfortunately
more engaged in speculating with the land,
than grazing upon or tilling it; and the consequence was, that in a few years the South
Australians were only saved from a famine by
the unexpected arrival overland of herds and
flocks from Victoria.   As it was, horses and
cows of a very indifferent kind were sold for

more than a hundred pounds a-piece, and sheep for five pounds a head.

The discovery of the copper mines alone saved the country from ruin. The first was the Kapunda. It was accidentally discovered by a shepherd, who picked up a piece on the surface of the ground, and showed it to his master. Pieces of copper ore may even now be found in the same way.

Next followed the far-famed Burra-Burra. In the latter mine there is a great quantity of malachite, which, when smelted, gives copper at an average of eighty-five per cent.

South Australia possesses the finest river in Australia—namely, the Murray, on which steamers will soon ply as far as five hundred miles up the country. On either side of this river is a thick and dry scrub—sometimes ten, sometimes thirty miles wide. In this scrub, manna is not unfrequently found, to the great delight of the natives, who are very fond of it. It is of a very excellent description, and in colour has a slight tendency to pink.

Adelaide itself is a well-laid out town. The streets are built in the same manner as in Sydney and Melbourne; but those in Adelaide are much wider. Many of the buildings and warehouses are highly creditable, particularly when we take the juvenile age of the colony into consideration.

Adelaide has never yet been "a transportation colony," and the society there is usually considered more *recherché* than in any other city in Australia. The climate is very good, and the vine flourishes as in the south of France. The principal export of South Australia is copper, to which may be added some wool and tallow.

The roads about are excellent, and the small farms in the neighbourhood are more in the English style than one could expect to meet with so many thousand miles away from home.

The overland route from Adelaide to Melbourne is about four hundred miles in length. In summer the road is pretty good, but in winter, a lake or swamp of twenty miles extent has to be waded through.

The scrub about South Australia is very thick, and any one may easily lose themselves in it. This has in fact often been the case, and a fearful instance of it occurred some few years ago. A young lady—the daughter of a gentleman residing near Adelaide—started out one Sunday afternoon with a book as her companion. Evening came, and she did not return, which alarmed her family, and search was made far and near—but in vain. On the fourth day, they at length discovered her lying dead at the foot of a tree. The pages of her book were covered with sentences, pricked in with a pin, expressive of her sufferings and of her unavailing efforts to retrace her steps. She was only three miles from her father's house when she sank down to die of hunger, thirst, and exhaustion ; and probably during the whole time of her wanderings had never exceeded that distance from her home.

The Adelaide gold-diggings began to excite attention in the months of August and September, 1852. In October the following report was made :

" Camp, Echunga Gold-Fields,
" October 2, 1852.

" Sir,

" I have the honour to state for the information of his Excellency the Lieutenant-Governor, that since my last report sixty licences have been issued, making a total of three hundred and fifty-six. * * * * Many families of respectability have arrived, and are now living in comfortable and commodious tents. The presence of well-dressed women and children gives to the gold-fields, apparently distinguished for decorum, security and respectability.

" From the feeling of greater security and comfort, combined with cheapness of living, all classes of diggers are unanimous in their preference of this place to Victoria. * * * *

" The nugget of gold which I have forwarded for his Excellency's inspection, weighing about an ounce and a half, was found about seven feet below the surface. * * * *

" There are some few amongst the lately arrived who expressed dissatisfaction with the result of their labours and observations, while others, who have been working for the last month, have promptly renewed their expired licences.

(Signed) "A. J. Murray,
"Assistant Gold Commissioner.

"The Hon. the Colonial Secretary."

In the month of October several pieces of gold, weighing each half an ounce and upwards, were found, and a few of the holes that had been abandoned by inexperienced hands, when taken possession of by old diggers on the Turon or the Bendigo, were found to contain good washing stuff. The diggings were well supplied with food of every kind; and during the summer months there could be no lack of fruits and vegetables in abundance, at reasonable prices, supplied from the numerous and well-cultivated farms and gardens around. This is certainly an advantage over the diggings of Victoria or New South Wales, if gold really does exist in paying quantities; if not, all the fruit and vegetables in the world would not keep the diggers at Echunga.

The following "Lament" was circulated in Adelaide, but not one of the newspapers there would print it. They were all too anxious for the success of their diggings, to countenance any grumblers against them :

A LAMENT FOR MY THIRTY SHILLINGS.

DEDICATED TO THE ECHUNGA VICTIMS.

My one pound ten! my one pound ten!
  I paid as Licence Fee ;
Ah! cruel Bonney! pray return
  That one pound ten to me.

When to Echunga diggings first
  I hastened up from town,
Thy tent I sought with anxious care
  And paid the money down.

And though my folly ever since
  I bitterly deplore,
It soothes my mind to know there were
  Three scores of fools before.

Then, Bonney, listen to my lay,
  And if you wish to thrive,

Send back the money quick to me,
  To number sixty-five.

Who wants but little here below,
  Nor wants that little long,
Had better to Echunga go,
  And not to Mount Coorong.

But as for me I like a swag,
  At least a little more
Than what we got there in a week—
  Eight pennyweights 'mongst four.

For that, of surface earth we washed
  Of dray loads half a score;
I'll swear that cradling never seemed
  Such tedious work before.

To sink for gold we then commenced,
  With grief I must confess,
'Twas fruitless toil, although we went
  Down thirty feet or less.

All you who've paid your one pound ten,
  Are on your licence told
That then you are entitled to
  Remove alluvial gold.

But if the alluvial gold's not there
  I'd like to have it proved

By what ingenious process it
   Can ever be removed?

Then back to Bendigo I'll haste,
   To seek the precious ore;
Although my one pound ten I fear
   Returns to me no more.

Yet as the boundary line I cross,
   My parting prayer shall be—
Ah! cruel Bonney! pray return
   My one pound ten to me!

ANTIGROPOLOS.

Adelaide, September 1852.

With a short extract from the "South Australian Register" of February 7, 1853, I shall conclude my remarks on the Adelaide diggings.

"*The Gold Fields.*—Although there is at the diggings everything to indicate gold in large quantities, none have succeeded in realizing their hopes. The majority content themselves with what they can get on Chapman's Hill and Gully, knowing that, if a fresh place is discovered, they will stand as good a chance as those who have spent months in trying to find better ground.

"The quantity of gold taken to the Assay-office, during four consecutive weeks, amounting to less than four thousand ounces, the Governor has proclaimed that after the 17th of February the office will be closed."

CHAPTER XVI.

MELBOURNE AGAIN.

It was on Monday the 25th of October, that for the second time I entered Melbourne. Not many weeks had elapsed since I had quitted it for my adventurous trip to the diggings, yet in that short space of time how many changes had taken place. The cloudy sky was exchanged for a brilliant sunshine, the chilling air for a truly tropical heat, the drizzling rain for clouds of thick cutting dust, sometimes as thick as a London fog, which penetrated the most substantial veil, and made our skins smart terribly. The streets too had undergone a wondrous transformation. Collins Street looked quite bright and cheerful, and was the fashionable promenade of those who had time or inclination for lounging. Parties of diggers were constantly starting or arriving, trips to St. Kilda and Brighton were daily taking place; and a coach was advertised to run to the diggings! I cannot quite realize the terrified passengers being driven through the Black Forest, but can picture their horror when ordered to " bail up" by a party of Australian Turpins.

In every window—milliners, baby-linen warehouses, &c., included—was exhibited the usual advertisement of the gold buyer—namely, a heap of gold in the centre, on one side a pile of sovereigns, on the other bank-notes. The most significant advertisement was one I saw in a window in Collins Street. In the middle was a skull perforated by a bullet, which lay at a little distance as if coolly examining or speculating on the mischief it had done. On one side of the skull was a revolver, and on the other a quantity of nuggets. Above all, was the emphatic inscription, " Beware in time."

This rather uncomfortable-looking tableau signified—in as speaking a manner as symbols
can—that the unfortunate skull had once belonged to some more unfortunate lucky digger,
who not having had the sense to sell his gold
to the proprietor of this attractive window had
kept his nuggets in his pocket, thereby tempt-
ing some robbers—significantly personified by
the revolver—to shoot him, and steal the gold.
Nowhere could you turn your eye without meeting
" 30,000 oz. wanted immediately ; highest price
given." " 10,000 oz. want to consign per——— ;
extra price given to immediate sellers," &c.
Outwardly it seemed a city of gold, yet hundreds were half perishing for want of food, with
no place of shelter beneath which to lay their
heads.   Many families of freshly-arrived emigrants—wife, children, and all—slept out in the
open air ; infants were born upon the wharves
with no helping hand near to support the
wretched mother in her misery.

How greatly the last few weeks had enlarged
Melbourne.   Cities of tents encompassed it on
all sides ; though, as I said before, the trifling

comfort of a canvas roof above them, was denied to the poorest of the poor, unless a weekly tax were paid!

But I must return to ourselves. Our first business the next morning was to find for our little Jessie some permanent home; for all our movements were so uncertain—I myself, thinking of a return to the old country—that it was considered advisable to obtain for her some better friends than a set of volatile, though good-hearted young fellows—not the most suitable protection for a young girl, even in so lax a place as the colonies. We never thought of letting her return to England, for there the life of a female, who has her own livelihood to earn, is one of badly-paid labour, entailing constant privation, and often great misery—if not worse. I have before said that William had relatives in Melbourne, and to them we determined to entrust her. Mrs. R—— was a kind-hearted and most exemplary woman; and having a very young family of her own, was well pleased at such an acquisition as the thoughtful, industrious little Jessie.

Each of our party contributed a small portion of their golden earnings to form a fund for a future day, which I doubt not will be increased by our little friend's industry, long before she needs it.   Here let us leave her, trusting that her future life may be as happy as her many excellent qualities deserve, and hoping that her severest trials have already passed over her.

Our next care was to obtain our gold from the Escort-office; to do which the receipts given in Bendigo had to be handed in, and after very little delay the precious packets were restored to their respective owners.   The following is a fac-simile of the tickets, printed on parchment, attached to each parcel of which a duplicate, printed on common paper, is given to the depositor :

BENDIGO CREEK.

No. 2772.

Date, 8th of October, 1852.

Name, Mr. A——.

Quantity, 60 oz. 10 dwts.

Consigned to, Self.

The trifling charge for all this trouble and responsibility is sixpence an ounce.

The business satisfactorily arranged, the next was to dispose of it. Some was converted into money, and sold for 69s. an ounce; and the remainder was consigned to England, where, being very pure and above standard, it realized £4 an ounce. A great difference that!

We next paid Richard a visit, who, though surprised, was well pleased to see us again. He declared his resolution of returning to England as soon as possible. Our party fixed their journey to the Ovens to take place in three weeks. William determined to remain in town, which I think showed wisdom on his part, as his health was not equal to roughing it in the bush; and this was a much more formidable trip than the last, on account of its greater length, and being much less frequented.

Meanwhile, we enjoyed the fine weather, and our present companionship, as much as possible, taking little trips here, there, and everywhere.

The one I most enjoyed was a sail in the Bay. The captain of the vessel in which we left England, was still detained in Port Philip for want of hands—the case of hundreds—and offered to give us a sail, and a dinner on board afterwards. We soon made up a large party, and enjoyed it exceedingly. The day was lovely. We walked down to Liardet's Beach, a distance of nearly three miles, and were soon calmly skimming over the waters. We passed St. Kilda and Brighton, and gained an excellent view of the innumerable vessels then lying useless and half-deserted in the Bay.

It was a sad though a pretty sight. There were fine East Indiamen, emigrant ships, American clippers, steamers, traders—foreign and English—whalers, &c., waiting there only through want of seamen.

In the cool of the evening our gallant host rowed us back to the beach. Since our first landing, tents and stores had been erected in great numbers, and Little Adelaide was grown wonderfully. I think I have never mentioned the quantity of frogs that abound in Australia.

This particular evening I remarked them more than usual, and without the least exaggeration their croaking resembled a number of mills in motion. I know nothing to which I can more appropriately liken the noise that resounded along the swampy portions of the road, from the beach to Melbourne.

Much has been said of the climate of Australia, and many are the conflicting statements thereon. The following table contains all the information—personal and otherwise—which I have been enabled to collect.

January and February.—Generally the hottest months; average of the thermometer, 78° in the shade; thunder-storms and *colonial* showers of rain occasionally visit us.

March.—Fine genial weather; average temperature, 73° in the shade.

April.—Weather more uncertain; mosquitos depart; average temperature, 70° in the shade.

May.—Fine, till towards the latter part of the month, when sometimes the rainy season commences; average temperature in the shade, 64°.

M

June.—Rainy, and much cooler; temperature at an average of 58° in the shade.

July.—Coldest month in the year; mid-winter in the colonies; average temperature, 53°. Ice and snow may be seen inland.

August.—Very rainy. Average temperature, 58° in the shade.

September.—Windy stormy month; weather getting warmer. Average temperature, 63° in the shade.

October.—The presence of the mosquito, a sure proof that the weather is permanently warm. Average temperature in the shade, 66°.

November and December.—Tropically warm. Locusts, mosquitos, and unnumbered creeping things swarm both in bush and town. Towards the end of December the creeks commence to dry up, and the earth looks parched for want of rain. No yule-log needed on Christmas Day. Thermometer as high as 97° in the shade; average 75°.

The principal trees in Australia are the gum, stringy bark, manna tree, wild cherry (so called),

iron bark, shea oak, peppermint, acacia, and the
mimosa, which last, however, should more
properly be called a shrub. These and others,
like the Indian maleleucas, are remarkable for the
Cajeput oil contained in their leaves, and in
the gums which exude from their stems, and in
this point of view alone, considering their
boundless number, their value can hardly be
over estimated. The gum of some of the
acacias will bear comparison with gum-arabic.
Their bark and timber are likewise useful, and
when the gold fever has subsided, will become
valuable as exports.

Wild flowers there are in abundance, and
some exquisite specimens of ferns. For the
benefit of those better skilled in botany than
myself, I give the following list of Dr. Müller's
indigenous plants of Victoria. Correaochrolenca
and Phebalium Asteriscophorum, both with the
medical properties of the Bucco-bush, Eurybia
Rhodochaeta, E. Rugosa, E. Adenophylla, E.
Asterotristia, Sambucus, Gaudichaudiana, Pros-
tanthera Hirsuta, Pimelea axiflora (powerful

Surrogat of the Mezerion shrub), Bossidea de-
cumbeus, Asterotristia asperifolia, Patersonia
aspera, Grevilliea repens, Dallachiana, &c.

The geranium, fuschia, rhododendrum, and
almost all varieties of the Cacti have been taken
to the colonies, and flourish well in the open air
all the year round, growing much more luxu-
riantly than in England.

The vineyards must some day form a considerable source of employment and profit to
the colonists.   The wine made in Australia is
very good.   The vines are cultivated in the
same manner as in France.   In the neighbourhood of Sydney, oranges and peaches are grown
out in the open air.   Apples and other fruits
flourish well in Van Diemen's Land.   All these
fruits are not indigenous to Australia.   The
only articles of food natural there, are the
kangaroos, emus, opossums, and other denizens
of the forest, a few snakes, some roots, and a
worm, about the length and thickness of a
finger, which is abundant in all parts of the
colony, and is taken out of the cavities, or from
under the bark, of the trees.   It is a great

favourite with the blacks, as it can be procured when no other food is attainable.

I have before made mention of the bush and scrub; there is a great dissimilarity between the two. The former resembles a forest, with none or very little underwood. The scrub, on the contrary, is always underwood, of from six to twenty feet high, and only here and there a few trees are seen. To be lost in either bush or scrub is a common thing. If on horseback the best way is to give the rein to your four-footed companion, and instinct will most probably enable him to extricate you. If on foot, ascend, if possible, a rise of ground, and notice any *fall* in the country; here, most likely, is a creek, and once beside that, you are pretty sure of coming to a station. If this fails, you must just bush it for the night, and resume your search next morning, trusting to an occasional " coo-ey" to help you out of your difficulty.

The scenery of Australia partakes of all characters. Sometimes miles of swamp reminds one of the Lincolnshire fens; at other times it assumes quite a park-like appearance, though

the effect is greatly injured by the want of
freshness about the foliage, which always looks
of a dirty, dingy green. The native trees in
Australia never shed their leaves, never have
that exquisite young tint which makes an English
spring in the country so delicious. Their faded
look always reminded me of those unfortunate
trees imprisoned for so many months beneath
the Crystal Palace.

The mountains in Australia are high and
bold in outline, and the snow-capped Alps on
the boundaries of New South Wales are not
unlike their European namesakes, the highest
tops are from six to seven thousand feet above
the level of the sea. The country round Ballarat
is more in the North American style, and when
the creek is full, it is a fine sight, greatly resembling, I have heard, one of the smaller rivers
in Canada; in fact, the scenery round Ballarat
is said to approach more to Upper Canada than
any in the colony. The rocks, although not
high, are in places very bold and romantic, and
in the wet season there are several water-falls
in the neighbourhood.

Eels are very plentiful in Victoria, and are peculiar to this district, being seldom, if ever, found in any other part of the known continent. Old writers on Australia have stated that eels are unknown in this part of the world, which, since this colony has been settled in, has been found to be erroneous, as the Barwin, the Yarra Yarra, and their tributaries abound with them, some weighing five or six pounds. A few days after our return from the diggings, we breakfasted off a dish of stewed eels, caught by a friend; the smallest weighed about a pound and a half, the largest about three pounds. They were caught three miles from Melbourne, in the Salt Water Creek.

A small kind of fish like the lamprey, another similar to the gudgeon, and also one (of rather a larger kind—the size of the roach) called here "white herrings," but not at all resembling that fish, are found. Pike are also very numerous. Crabs and lobsters are not known here, but in the salt creeks near the sea we have craw-fish.

Of course, parrots, cockatoos and "sich-like,"

abound in the bush, to the horror of the small gardeners and cultivators, as what they do not eat they ruin by destroying the young shoots.

Kangaroos are extremely numerous in the scrub. They are the size of a large greyhound, and of a mouse colour. The natives call them "kanguru." The tail is of great strength. There are several varieties of them. The largest is the Great Kangaroo, of a greyish-brown colour, generally four or five feet high and the tail three. Some kangaroos are nearly white, others resemble the hare in colour. Pugs, or young kangaroos, are plentiful about the marshy grounds; so are also the opossum and kangaroo rat. The latter is not a rat, properly speaking, but approaches the squirrel tribe. It is a lilliputian kangaroo, the size of our native wood-squirrel, and larger, only grey or reddish-grey. It can leap six or eight feet easily, and is excellent eating. The native dog is of all colours; it has the head and brush of a fox, with the body and legs of a dog. It is a cowardly animal, and will run away from you like mad. It is a great enemy of the kangaroo rat, and a torment to the squatter, for a native dog has a great *penchant* for mutton, and will kill thirty or forty sheep in the course of an hour.

A species of mocking-bird which inhabits the bush is a ludicrous creature. It imitates everything, and makes many a camping party imagine there is a man near them, when they hear its whistle or hearty laugh. This bird is nicknamed the " Jackass," and its loud " ha ! ha ! ha !" is heard every morning at dawn echoing through the woods and serving the purpose of a "boots" by calling the sleepy traveller in good time to get his breakfast and pursue his journey. The bats here are very large.

Insects, fleas, &c., are as plentiful as it is possible to be, and the ants, of which there are several kinds, are a perfect nuisance. The largest are called by the old colonists, " bull-dogs," and formidable creatures they are— luckily not very common, about an inch and a half long, black, or rusty-black, with a red tail. They bite like a little crab. Ants of an inch long are quite common. They do not— like the English ones—run scared away at the sight of a human being — not a bit of it; Australian ants have more *pluck*, and will turn and face you. Nay, more, should *you* retreat, they will run after you with all the impudence imaginable. Often when my organ of destruc- tiveness has tempted me slightly to disturb with the end of my parasol one of the many ant-hills on the way from Melbourne to Rich- mond, I have been obliged, as soon as they discovered the perpetrator of the attack, to take to my heels and run away as if for my life.

Centipedes and triantelopes (colonial, for tarantula) are very common, and though not exactly fatal, are very dangerous if not attended to. The deaf adder is the most formidable " varmint " in Australia. There are two varie- ties; it is generally about two feet long. The bite is fatal. The deaf adder never moves unless it is touched, hence its name. I do not think it has the power of twisting or twirling, like the ordinary snake or adder and it is very slow in its movements. There are several species of snakes, some of them are extremely venomous and grow to a large size, as long as ten feet. The black snake is the most venomous of any; its bite is fatal within a few hours.

But let us leave these wilder subjects and return to Melbourne.

The state of society in the town had not much improved during my absence. On the public road from Melbourne to St. Kilda, fifteen men were robbed in one afternoon, and tied to trees within sight of one another. In Melbourne itself the same want of security prevailed, and concerts, lectures, &c., were always advertised to take place when there was a full moon, the only nights any one, unarmed, dared venture out after dusk. The following extract from the "Argus," gives a fair specimen of Melbourne order.

"We are led to these remarks (referring to a tirade against the Government) by an occurrence that took place last week in Queen Street, the whole detail of which is peculiarly illustrative of the very creditable state of things, to which, under the happy auspices of a La Trobe dynasty, we are rapidly descending.

" A ruffian robs a man in a public-house, in broad daylight. He is pursued by a constable and taken. On the way to the watchhouse a mob collects, the police are attacked, pistols are pointed, bludgeons and axe-handles are brought out of the adjacent houses (all still in broad daylight, and in a busy street), and distributed amongst the crowd, loud cries inciting attack are heard, a scuffle ensues, the police are beaten, the prisoner is rescued, the crowd separates, and a man is left dead upon the ground. The body is taken into a public-house, an inquest is held, the deceased is recognized as a drunkard, the jury is assured that a *post-mortem* examination is quite unnecessary; and the man is buried, after a verdict is brought in of 'Died by the visitation of God;' the said visitation of God having, in this instance, assumed the somewhat peculiar form of a fractured skull!"

This is a true picture of Melbourne; but

whether the "Argus" is justified in reproach-
ing the "La Trobe dynasty" with it, is quite
another matter.

In pages like these, anything resembling an
argument on the "transportation question,"
would be sadly out of place. To avoid thinking or hearing it was impossible, for during my
second stay in Melbourne, it was a never-failing
subject of conversation. In Victoria (which is
only forty-eight hours' journey from Van Diemen's Land), I have seen the bad results of the
mingling of so many transports and ticket-of-
leave men among the free population. On the
other hand, I have heard from many and good
authorities, of the substantial benefits conferred
on Sydney and New South Wales by convict
labour. It is difficult to reconcile these two
statements, and it is an apple of discord in the
colonies.

Whilst in Victoria, I met with a great variety
of emigrants, and I was much struck by the
great success that seems to have attended on
almost all of those who came out under the
auspices of Mrs. Chisholm. No one in Eng-

land can fully appreciate the benefits her un-
wearied exertions have conferred upon the
colonies. I have met many of the matrons of
her ships, and not only do they themselves
seem to have made their way in the world, but
the young females who were under their care
during the voyage appear to have done equally
well. Perhaps one way of accounting for this,
is the fact that a great many of those going out
by the Chisholm Society are from Scotland,
the inhabitants of which country are peculiarly
fortunate in the colonies, their industry, fru-
gality, and " canniness " being the very qualities
to make a fortune there. " Sydney Herbert's
needlewomen " bear but a bad name; and the
worst recommendation a young girl applying
for a situation can give, is to say she came out
in that manner—not because the colonists look
down on any one coming out by the assistance
of others, but because it is imagined her female
associates on the voyage cannot have been such
as to improve her morality, even if she were
good for anything before.

Much is said and written in England about

the scarcity of females in Australia, and the many good offers awaiting the acceptance of those who have the courage to travel so far. But the colonial bachelors, who are so ready to get married, and so very easy in their choice of a wife, are generally those the least calculated, in spite of their wealth, to make a respectable girl happy; whilst the better class of squatters and diggers—if they do not return home to get married, which is often the case—are not satisfied with any one, however pretty, for a wife, unless her manners are cultivated and her principles correct.

To wander through Melbourne and its environs, no one would imagine that females were as one to four of the male population; for bonnets and parasols everywhere outnumber the wide-awakes. This is occasioned by the absence of so many of the "lords of creation" in pursuit of what they value—many of them, at least—more than all the women in the world—nuggets. The wives thus left in town to deplore their husbands' infatuation, are termed "grass-widows"—a mining expression.

And now two out of the three weeks of our party's stay in Melbourne has expired, during which time a change (purely personal) had made my brother's protection no longer needed by me. *My* wedding-trip was to be to England, and the marriage was to take place, and myself and *caro sposo* to leave Australia before my brother departed for the Ovens diggings. The ' C——,' a fine East Indiaman, then lying in the bay, was bound for London. We were to be on board by the 12th of November.

This of course gave me plenty to do, and my last morning but one in Melbourne was dedicated to that favourite feminine occupation— which, however, I detest—I mean, shopping. This being accomplished to my great dissatisfaction—for all I bought could have been obtained, of a better description, for half the price in England—I was preparing to return home by way of Collins Street, when my name in familiar accents made me suddenly pause. I instantly recognised the lady who addressed me as one of the English governesses in a " finishing" school where three years of my girlhood were passed. Julia —— was a great favourite among us; no one could have done otherwise than admire the ability and good-humour with which she fulfilled her many arduous duties. Perhaps, of all miserable positions for a well-educated and refined young person to be placed in, that of "little girls' teacher" in a lady's school is the worst.

Her subsequent history I learnt as we walked together to my present abode.

Her mother had had a cousin in Sydney, who being old and unmarried, wrote to her, promising to settle all his property, which was considered large, upon her daughter and herself, his only living relatives, provided they came out to the colonies to live with him until his death. A sum of money to defray the expenses of the voyage was enclosed. This piece of un-expected good news was received with pleasure, and the invitation gladly accepted. They sailed for Sydney. On arriving there, they found that some speculation, in which he was greatly involved, had failed; and the old man had taken the loss so greatly to heart, that he died only five months after having dispatched the letter to his English relatives.

Poor Julia was placed in a most painful position. In England she had scarcely been able to support her invalid mother by her own exertions, but in a strange country and without friends these difficulties seemed increased. Her first act was to look over the advertizing columns of the papers, and her eye caught sight of one which seemed exactly to suit her. It was, "Wanted, a governess to take the entire charge of a little girl, the daughter of a widower, and also an elderly lady, to superintend the domestic arrangements of the same family during the continual absence of the master at another station." Julia wrote immediately, and was accepted. In the occasional visits that her pupil's father paid to his little girl, he could not fail to be struck by the sweet disposition and many other good qualities of her governess, and it ended by his making her his wife. I felt at liberty to congratulate her, for she looked the picture of happiness. I saw her again next day, when she showed me the advertisement which had been the means of such a change in her circumstances.

The day before my departure was a painful one, so many farewells to be taken of dear friends whom I should never meet again.

On Friday, the 15th of November, my brother and all our party, Richard and Jessie included, accompanied us to the pier at Williamstown, to which we were conveyed by a steamer. For this we paid five shillings a-piece, and the same for each separate box or parcel, and twelve shillings to a man for carting our luggage down to the Melbourne wharf, a distance of not a mile.

On landing at the pier, how greatly was I astonished to meet Harriette and her husband. Her modest desires were gratified, and they had realized sufficient capital at the diggings to enable them to settle most comfortably near Adelaide. In hurried words she told me this, for their boat was already alongside the pier waiting to take them to their ship. Hardly had they departed than a boat arrived from

our vessel to convey us to it. Sad adieux were
spoken, and we were rowed away.

That evening a pilot came on board, anchors
were weighed, we left the bay, and I saw Melbourne no more.

CHAPTER XVII.

HOMEWARD BOUND.

WE soon left Port Philip far behind, and in a few days saw nothing but a vast expanse of water all around us. Our vessel was filled with returning diggers; and it is worth while to remark that only two had been unsuccessful, and these had only been at the diggings a few days.

One family on board interested me very much. It consisted of father, mother, and two children. The eldest, a little girl, had been born some time before they left England. Her brother was a sturdy fellow of two years old,

born in the colonies soon after their arrival.
He could just toddle about the deck, where he
was everlastingly looking for "dold," and
"nuddets." The whole family had been at the
diggings for nine months, and were returning
with something more than £2,000 worth of
gold. In England it had been hard work to
obtain sufficient food by the most constant
labour; they had good reason to be thankful
for the discovery of the gold-fields.

Saturday, November 27, was forty-eight
hours long, or two days of the same name
and date. Sailing right round the world in
the direction of from west to east, we gained
exactly twenty-four hours upon those who stay
at home; and we were therefore obliged to
make one day double to prevent finding ourselves wrong in our dates and days on our
arrival in England. Melbourne is about ten
hours before London, and therefore night and
day are reversed.

Rapidly it became cooler, for the winds were
rather contrary, and drove us much farther
south than was needed. We were glad to avail

ourselves of our opossum rugs to keep ourselves
warm. One of these rugs is quite sufficient
covering of a night in the coldest weather, and
imparts as much heat as a dozen blankets.
They are made from the skins of the opossums,
sewn together by the natives with the sinews of
the same animal. Each skin is about twelve
inches by eight, or smaller; and as the rugs are
generally very large, they contain sometimes as
many as eighty skins. They may be tastefully
arranged, as there is a great difference in the
colours; some being like a rich sable, others
nearly black, and others again of a grey and
light brown. The fur is long and silky. At
one time a rug of this description was cheap
enough—perhaps as much as two sovereigns—
but the great demand for them by diggers, &c.,
has made them much more scarce, and it now
requires a ten pound-note to get a good one.
The best come from Van Diemen's Land. In
summer they are disagreeable, as they harbour
insects.

However, whilst rounding Cape Horn, in the
coldest weather I ever experienced, we were

only too happy to throw them over us during the nights.

One morning we were awakened by a great confusion on deck. Our ship was ploughing through a quantity of broken ice. That same afternoon, we sighted an immense iceberg about ten miles from us. Its size may be imagined from the fact, that, although we were sailing at a rate of ten knots an hour, we kept it in sight till dark. This was on the 3rd of December.

We soon rounded the Horn, and had some very rough weather. One of the sailors fell off the jib-boom ; and the cry of " man overboard" made our hearts beat with horror. Every sail was on ; we were running right before the wind, and the waves were mountains high, a boat must have been swamped ; and long before we could " 'bout ship," he had sunk to rise no more.

After rounding Cape Horn, we made rapid progress ; by Christmas Day, we were in the Tropics. It was not kept with much joviality, for water and food were running scarce. Pro-

visions were so dear in Melbourne, that they
had laid in a short allowance of everything, and
our captain had not anticipated half so many
passengers. We tried, therefore, to put into
St. Helena, but contrary winds preventing us,
we sailed back again to the South American
coast, and anchored off Pernambuco. It was
providential that economical intentions made
our captain prefer this port, for had we touched
at Rio, we should have encountered the yellow
fever, which we afterwards heard was raging
there.

Pernambuco is apparently a very pretty
place. We were anchored about four miles
from the town, so had a good view of the coast.
I longed to be on shore to ramble beneath the
elegant cocoa-nut-trees. The weather was intensely hot, for it was in the commencement
of January; and the boats full of fruit, sent
from the shore for sale, were soon emptied by
us. I call them boats, but they are properly
termed catamarans. They are made of logs of
wood lashed securely together; they have a sail
and oars but no sides, so the water rushes over,

N

and threatens every moment to engulf the frail conveyance; but no, the wood is too light for that. The fruits brought us from shore were oranges, pine-apples, water-melons, limes, bananas, cocoa-nuts, &c., and some yams, which were a good substitute for potatoes. The fruit was all very good, and astonishingly cheap; our oranges being green, lasted till we reached England. Some of our passengers went on shore, and returned with marvellous accounts of the dirtiness and narrowness of the streets, and the extremely *natural* costume of the natives.

We remained here about four days, and then, with favourable winds, pursued our voyage at an average rate of ten or twelve knots an hour. As we neared the English coast, our excitement increased to an awful height; and for those who had been many years away, I can imagine every trivial delay was fraught with anxiety.

But we come in sight of land; and in spite of the cold weather, for it is now February, 1853, every one rushes to the deck. On we go; at last we are in the Downs, and then the wind turned right against us.

Boats were put off from the Deal beach. The boatmen there rightly calculated that lucky gold-diggers wouldn't mind paying a pound a-piece to get ashore, so they charged that, and got plenty of customers notwithstanding.

On Sunday, the 27th of February, I again set foot on my native land. It was evening when we reached the shore, and there was only an open vehicle to convey us to the town of Deal itself. The evening was bitterly cold, and the snow lay upon the ground. As we proceeded along, the sounds of the Sabbath bell broke softly on the air. No greeting could have been more pleasing or more congenial to my feelings.

CHAPTER XVIII.

CONCLUSION.

As I trust that, in the foregoing pages, I have slightly interested my readers in " our party," the following additional account of their movements, contained in letters addressed to me by my brother, may not be quite uninteresting.

The Ovens diggings are on the river of the same name, which takes its rise in the Australian Alps, and flows into the Murray. As these Alps separate New South Wales from Victoria, these diggings belong to the latter province. They are about forty miles from the town of Albury. They are spread over a large space of ground. The principal localities are Spring and Reid's Creeks.

Now for the letters.

"Melbourne, January 5, 1853.

"My dear E—,

"You'll be surprised at the heading of this; but the Ovens are not to my taste, and I'm off again with Frank and Octavius to Bendigo to-morrow. I suppose you'll like to hear of our adventures up to the Ovens, and the reasons for this sudden change of plans. We left Melbourne the Monday after you sailed, and camped out half-way to Kilmore, a little beyond the 'Lady of the Lake.' The day was fine, but the dust made us wretched. Next day, we reached Kilmore—stopped there all night. Next day on again, and the farther we went, the more uncivilized it became—hills here, forests there, as wild and savage as any one could desire. It was 'bushing it' with a vengeance. This lasted several days. Once we

lost our road, and came, by good luck, to a sort
of station. They received us very hospitably,
and set us right next morning. Four days
after, we came to the Goulburn river. There
was a punt to take us over, and a host of
people (many from Bendigo) waiting to cross.
Three days after, we pitched out tents at the
Ovens. Here I soon saw it was no go. There
was too much water, and too little gold; and
even if they could knock the first difficulty on
the head, I don't think they could do the same
to the second. In my own mind, I think it
impossible that the Ovens will ever turn out
the second Bendigo that many imagine. Hundreds differ from me, therefore it's hundreds to
one that I'm wrong. The average wages, as
far as I can judge, are an ounce a-week; some
much more, many much less. We did not
attempt digging ourselves. Eagle Hawk shal-
lowness has spoilt us, for not even Octavius
(who, you know of old, was a harder worker
than either Frank or self,) thinks it worth
digging through fourteen or sixteen feet of

hard clay for the mere pleasure of exercising our limbs. Provisions there were not at the high price that many supposed they would be, but quite high enough, Heaven knows! Meat was very scarce and bad, and flour all but a shilling a pound; and if the fresh arrivals keep flocking in, and no greater supply of food, it will get higher still. We stayed there two weeks, then brought our dray back again, and are now busy getting ready for a fresh start to Bendigo. Among other things we shall take, are lemonade and ginger-beer powders, a profitable investment, though laughable. The weather is very hot —fancy 103° in the shade. Water is getting scarce.

*　　　*　　　*　　　*

" Have seen all our friends in Melbourne except Richard, who left for England a fortnight ago. Jessie is well, and growing quite pretty. She says she is extremely happy, and sends such a number of messages to you, that I'll write none, for fear of making a mistake. Will write again soon.

*   *   *   *

" Your affectionate brother, in haste,

"_____."

" Melbourne, April 17, 1853.

" My dear E—,

" I suppose you've thought I was buried in my hole, or 'kilt' by bushrangers in the Black Forest; but I've been so occupied in the worship of Mammon, as to have little thoughts for anything else.

*   *   *   *

" We made a good thing of our last two speculations. Ginger-beer and lemonade, or lemon kali, at sixpence a tiny glass, paid well. A successful digger would drink off a dozen one after another. Some days, we have taken ten pounds in sixpences at this fun. What they bought of us wouldn't harm them, but many mix up all sorts of injurious articles to sell; but our consciences, thank God! are not colonised sufficiently for that. We have had steady good luck in the digging line (for we combine everything), and, after this next trip, mean to dissolve partnership.

*   *   *   *

" Octavius talks of going out as overseer, or something of that sort, to some squatter in New South Wales for a year or so, just to learn the system, &c., and then, if possible, take a sheep-run himself. Frank means to send for Mrs. Frank and small Co. He says he shall stay in Victoria for some years. I do believe he likes the colony. As for myself, I hope to see the last of it in six weeks' time.

*   *   *   *

" Hurrah for Old England !—no place like it.

*   *   *   *

" Your very affectionate brother,

" _____."

With a cordial assent to the last few words, I conclude these pages.

APPENDIX.

WHO SHOULD EMIGRATE?

THE question of "Who should emigrate?" has now become one of such importance (owing to the number who are desirous of quitting their native land to seek a surer means of subsistence in our vast colonial possessions), that any book treating of Australia would be sadly deficient were a subject of such universal interest to be left unnoticed; and where there are so many of various capabilities, means and dispositions, in need of guidance and advice as to the advantage of their emigrating, it is probable that the experience of any one, however slight that experience may be, will be useful to some.

Any one to succeed in the colonies must take with him a quantity of self-reliance, energy, and perseverance ; this is the best capital a man can have.   Let none rely upon introductions—they are but useless things at the best—they may get you invited to a good dinner ; but now that fresh arrivals in Melbourne are so much more numerous than heretofore, I almost doubt if they would do even that.   A quick, clever fellow with a trade of his own, inured to labour, and with a light heart, that can laugh at the many privations which the gipsy sort of life he must lead in the colonies will entail upon him ; any one of this description cannot fail to get on.   But for the number of clerks, shopmen, &c., who daily arrive in Australia, there is a worse chance of their gaining a livelihood than if they had remained at home. With this description of labour the colonial market is largely overstocked ; and it is distressing to notice the number of young men incapable of severe manual labour, who, with delicate health, and probably still more delicately filled purses, swarm the towns in search of employment, and are exposed to heavy expenses which they can earn nothing to meet. Such men have rarely been successful at the diggings; the demand for them in their accustomed pursuits is very limited in proportion to their numbers; they gradually sink into extreme poverty—too often into reckless or criminal habits — till they disappear from the streets to make way for others similarly unfortunate.

A little while since I met with the histories of two individuals belonging to two very different classes of emigrants; and they are so applicable to this subject, that I cannot forbear repeating them.

The first account is that of a gentleman who went to Melbourne some eight months ago, carrying with him a stock of elegant acquirements and accomplishments, but little capital. He is now in a starving condition, almost without the hope of extrication, and is imploring from his friends the means to return to England, if he live long enough to receive them. The colours in which he paints the colonies are deplorable in the extreme.

The other account is that of a compositor who emigrated much about the same time. He writes to his former office-mates that he got immediate and constant employment at the rate of £7 per week, and naturally thinks that there is no place under the sun like Melbourne. Both emigrants are right. There is no better place under the sun than Melbourne for those who can do precisely what the Melbourne people want; and which they must and will have at any price; but there is no worse colony to which those can go who have not the capabilities required by the Melbourne people. They are useless and in the way, their accomplishments are disregarded, their misfortunes receive no pity; and, whilst a good carpenter or bricklayer would make a fortune, a modern Raphael might starve.

But even those possessed of every qualifi-
cation for making first-class colonists, will at
first meet with much to surprise and annoy
them, and will need all the energy they possess,
to enable them to overcome the many disagreeables which encounter them as soon as
they arrive.

Let us, for example, suppose the case of
an emigrant, with no particular profession or
business, but having a strong constitution, good
common sense, and a determination to bear
up against every hardship, so that in the end
it leads him to independence. Let us follow
him through the difficulties that bewilder the
stranger in Melbourne during the first few
days of his arrival.

The commencement of his dilemmas will
be that of getting his luggage from the ship;
and so quickly do the demands for pounds and
shillings fall upon him, that he is ready to
wish he had pitched half his "traps" overboard. However, we will suppose him at
length safely landed on the wharf at Melbourne,

with all his boxes beside him.   He inquires
for a store, and learns that there are plenty
close at hand; and then forgetting that he is
in the colonies, he looks round for a porter and
truck, and looks in vain.   After waiting as
patiently as he can for about a couple of hours,
he manages to hire an empty cart and driver;
the latter lifts the boxes into the conveyance
(expecting, of course, his employer to lend a
hand), smacks his whip, and turns down street
after street till he reaches a tall, grim-looking
building, in front of which he stops, with a
"That ere's a store," and a demand for a
sovereign, more or less.   This settled, he coolly
requests the emigrant to assist him in unloading,
and leaves him to get his boxes carried inside
as best he can.   Perhaps some of the store-
keeper's men come to the rescue, and with
their help the luggage is conveyed into the
store-room (which is often sixty or eighty
feet in length), where the owner receives a
memorandum of their arrival.   Boxes or
parcels may remain there in perfect safety

for months, so long as a shilling a week is
paid for each.

Our emigrant, having left his property in
security, now turns to seek a lodging for himself; and the extreme difficulty of procuring
house accommodation, with its natural consequences, an extraordinary rate of rent, startles
and amazes him. He searches the city in
vain, and betakes himself to the suburbs,
where he procures a small, half-furnished room,
in a wooden house for thirty shillings a week.
The scarcity of houses in proportion to the
population, is one of the greatest drawbacks
to the colony; but we could not expect it to be
otherwise when we remember that in one year
Victoria received an addition of nearly 80,000
inhabitants. The masculine portion of these
emigrants, with few exceptions, started off at
once to the diggings; hence the deficiency in
the labour market is only partially filled up by
the few who remained behind, and by the
fewer still who forsake the gold-fields ; whilst the
abundance of money, and the deficiency of good

workmen, have raised the expenses of building far above the point at which it would be a profitable investment for capital.   Meantime, the want is only partially supplied by the wooden cottages which are daily springing up around the boundaries of the city ; but this is insufficient to meet the increasing want of shelter, and on the southern bank of the Yarra there are four or five thousand people living in tents.   This settlement is appropriately called "Canvas Town."

But let us return to our newly-arrived emigrant.

Having succeeded in obtaining a lodging, he proceeds to purchase some necessary articles of food, and looks incredulously at the shopkeeper when told that butter is 3s. 6d. a pound, cheese, ham, or bacon 2s. to 2s. 6d., and eggs 4s. or 5s. a dozen.   He wisely dispenses with such luxuries, and contents himself with bread at 1s. 6d. the four-pound loaf, and meat at 5d. a pound.   He sleeps soundly, for the day has been a fatiguing one, and next morning with

renewed spirits determines to search immediately for employment. He does not much care
what it is at first, so that he earns something;
for his purse feels considerably lighter after the
many demands upon it yesterday. Before an
hour is over, he finds himself engaged to a
storekeeper at a rate of £3 a-week; his business
being to load and unload drays, roll casks, lift
heavy goods, &c.; and here we will leave him,
for once set going he will soon find a better
berth. If he have capital, it is doubtless safely
deposited in the Bank until a little acquaintance
with the colonies enables him to invest it judiciously; and meanwhile, if wise, he will spend
every shilling as though it were his last; but
if his capital consists only of the trifle in his
purse, no matter, the way he is setting to
work will soon rectify that deficiency, and
he stands a good chance in a few years of
returning to England a comparatively wealthy
man.

To those of my own sex who desire to emigrate to Australia, I say do so by all means, if

you can go under suitable protection, possess
good health, are not fastidious or "fine-lady-
like," can milk cows, churn butter, cook a good
damper, and mix a pudding.   The worst risk
you run is that of getting married, and finding
yourself treated with twenty times the respect
and consideration you may meet with in England.   Here (as far as number goes) women
beat the "lords of creation ;" in Australia it is
the reverse, and there we may be pretty sure
of having our own way.

But to those ladies who cannot wait upon themselves, and whose fair fingers are unused to the
exertion of doing anything useful, my advice is,
for your own sakes remain at home.   Rich
or poor, it is all the same ; for those who can
afford to give £40 a-year to a female servant
will scarcely know whether to be pleased or not
at the acquisition, so idle and impertinent are
they ; scold them, and they will tell you that
" next week Tom, or Bill, or Harry will be
back from the diggings, and then they'll be
married, and wear silk dresses, and be as fine a

lady as yourself;" and with some such words will coolly dismiss themselves from your service, leaving their poor unfortunate mistress uncertain whether to be glad of their departure or ready to cry because there's nothing prepared for dinner, and she knows not what to set about first.

For those who wish to invest small sums in goods for Australia, boots and shoes, cutlery, flash jewellery, watches, pistols (particulary revolvers), gunpowder, fancy articles, cheap laces, and baby-linen offer immense profits.

The police in Victoria is very inefficient, both in the towns and on the roads. Fifteen persons were stopped during the same afternoon whilst travelling on the highway between Melbourne and St. Kilda. They were robbed, and tied to trees within sight of each other—this too in broad daylight. On the roads to the diggings it is still worse; and no one intending to turn digger should leave England without a good supply of fire-arms In less than one

week more than a dozen robberies occurred
between Kyneton and Forest Creek, two of
which terminated in murder.   The diggings
themselves are comparatively safe — quite as
much so as Melbourne itself—and there is a
freemasonry in the bush which possesses an
irresistible charm for adventurous bachelors,
and causes them to prefer the risk of bushrangers to witnessing the dreadful scenes that
are daily and hourly enacting in a colonial town.
Life in the bush is wild, free and independent.
Healthy exercise, fine scenery, and a clear and
buoyant atmosphere, maintain an excitement of
the spirits and a sanguineness of temperament
peculiar to this sort of existence ; and many are
the pleasant evenings, enlivened with the gay
jest or cheerful 'song, which are passed around
the bush fires of Australia.

The latest accounts from the diggings speak
of them most encouragingly.   Out of a population of 200,000 (which is calculated to be
the number of the present inhabitants of Victoria), half are said to be at the gold-fields, and

the average earnings are still reckoned at nearly an ounce per man per week. Ballarat is again rising into favour, and its riches are being more fully developed. The gold there is more un-equally distributed than at Mount Alexander, and therefore the proportion of successful to unsuccessful diggers is not so great as at the latter place. But then the individual gains are in some cases greater. The labour is also more severe than at the Mount or Bendigo, as the gold lies deeper, and more numerous trials have to be made before the deposits are struck upon.

The Ovens is admitted to be a rich gold-field, but the work there is severely laborious, owing to a super-abundance of water.

The astonishing mineral wealth of Mount Alexander is evidenced by the large amounts which it continues to yield, notwithstanding the immense quantities that have already been taken from it. The whole country in that neighbourhood appears to be more or less auriferous.

Up to the close of last year the total supposed amount of gold procured from the Victoria diggings, is 3,998,324 ounces, which, when calculated at the average English value of £4 an ounce, is worth nearly *sixteen millions sterling*. One-third of this is distinctly authenticated as having come down by escort during the three last months of 1852.

In Melbourne the extremes of wealth and poverty meet, and many are the anecdotes of the lavish expenditure of successful diggers that are circulated throughout the town. I shall only relate two which fell under my own observation.

Having occasion to make a few purchases in the linendrapery line, I entered a good establishment in Collins Street for that purpose. It was before noon, for later in the day the shops are so full that to get a trifling order attended to would be almost a miracle. There was only one customer in the shop, who was standing beside the counter, gazing with extreme dissatisfaction upon a quantity of goods of various colours and materials that lay there for his inspection. He was a rough-looking customer enough, and the appearance of his hands gave strong indication that the pick-axe and spade were among the last tools he had handled.

" It's a *shiny* thing that I want," he was saying as I entered.

" These are what we should call shining goods," said the shopman, as he held up the silks, alpacas, &c., to the light.

" They're not the *shiny* sort that I want," pursued the digger, half-doggedly, half-angrily. " I'll find another shop; I guess you won't show your best goods to me—you think, mayhap, I can't pay for them—but I can, though," and he laid a note for fifty pounds upon the counter, adding, " maybe you'll show me some *shiny* stuff now."

Unable to comprehend the wishes of his customer, the shopman called to his assistance the master of the establishment, who being, I suppose, of quicker apprehension, placed some satins before him.

" I thought the paper would help you find it.
I want a gown for my missus. What's the
price?"

" Twenty yards at one-ten—thirty pounds.
That do, Sir?"

" No; not good enough!" was the energetic
reply.

The shrewd shopkeeper quickly fathomed his
customer's desires, and now displayed before
him a rich orange-coloured satin, which elicited
an exclamation of delight.

" Twenty-five yards—couldn't sell less, it's a
remnant—at three pounds the yard."

" That's the go!" interrupted the digger,
throwing some more notes upon the counter. " My missus was married in a cotton
gown, and now she'll have a real gold
'un!"

And seizing the satin from the shopkeeper,
he twisted up the portion that had been
unrolled for his inspection, placed the whole
under his arm, and triumphantly walked out
of the shop, little thinking how he had been
cheated.

" A 'lucky digger' that," observed the shop-man, as he attended to my wants.

I could not forbear a smile, for I pictured to myself the digger's wife mixing a damper with the sleeves of her dazzling satin dress tucked up above her elbows.

A few days after, a heavy shower drove me to take shelter in a pastry-cook's, where, under the pretence of eating a bun, I escaped a good drenching.   Hardly had I been seated five minutes, when a sailor-looking personage en-tered, and addressed the shopwoman with: " I'm a-going to be spliced to-morrow, young woman ; show us some large wedding-cakes."

The largest (which was but a small one) was placed before him, and eighteen pounds de-manded for it.   He laid down four five-pound notes upon the counter, and taking up the cake, told her to " keep the change to buy ribbons with."

" Pleasant to have plenty of gold-digging friends," I remarked, by way of saying some-thing.

" Not a friend," said she, smiling. " I never saw him before. I expect he's only a successful digger."

Turn we now to the darker side of this picture.

My favourite walk, whilst in Melbourne, was over Prince's Bridge, and along the road to Liardet's Beach, thus passing close to the canvas settlement, called Little Adelaide. One day, about a week before we embarked for England, I took my accustomed walk in this direction, and as I passed the tents, was much struck by the appearance of a little girl, who, with a large pitcher in her arms, came to procure some water from a small stream beside the road. Her dress, though clean and neat, bespoke extreme poverty; and her countenance had a wan, sad expression upon it which would have touched the most indifferent beholder, and left an impression deeper even than that produced by her extreme though delicate beauty.

I made a slight attempt at acquaintanceship

by assisting to fill her pitcher, which was far
too heavy, when full of water, for so slight a
child to carry, and pointing to the rise of
ground on which the tents stood, I inquired if
she lived among them.

She nodded her head in token of assent.

" And have you been long here? and do you
like this new country ?" I continued, determined to hear if her voice was as pleasing as
her countenance.

" No !" she answered quickly ; " we starve
here.   There was plenty of food when we
were in England;" and then her childish
reserve giving way, she spoke more fully of
her troubles, and a sad though a common
tale it was.

Some of the particulars I learnt afterwards.
Her father had held an appointment under
Government, and had lived upon the income
derived from it for some years, when he was
tempted to try and do better in the colonies.
His wife (the daughter of a clergyman, well
educated, and who before her marriage had been

a governess) accompanied him with their three children.   On arriving in Melbourne (which was about three months previous), he found that situations equal in value, according to the relative prices of food and lodging, to that which he had thrown up in England were not so easily procured as he had been led to expect. Half desperate, he went to the diggings, leaving his wife with little money, and many promises of quick remittances of gold by the escort.   But week followed week, and neither remittances nor letters came.   They removed to humbler lodgings, every little article of value was gradually sold, for, unused to bodily labour, or even to sit for hours at the needle, the deserted wife could earn but little.   Then sickness came ; there were no means of paying for medical advice, and one child died.   After this, step by step, they became poorer, until half a tent in Little Adelaide was the only refuge left.

As we reached it, the little girl drew aside the canvas, and partly invited me to enter.   I

glanced in ; it was a dismal sight. In one
corner lay the mother, a blanket her only protection from the humid soil, and cowering down
beside her was her other child. I could not
enter ; it seemed like a heartless intrusion upon
misery ; so, slipping the contents of my purse
(which were unfortunately only a few shillings)
into the little girl's hand, I hurried away,
scarcely waiting to notice the smile that
thanked me so eloquently. On arriving at
home, I found that my friends were absent,
and being detained by business, they did not
return till after dusk, so it was impossible for
that day to afford them any assistance. Early
next morning we took a little wine and other
trifling articles with us, and proceeded to Little
Adelaide. On entering the tent, we found that
the sorrows of the unfortunate mother were at
an end ; privation, ill-health, and anxiety had
claimed their victim. Her husband sat beside
the corpse, and the golden nuggets, which in
his despair he had flung upon the ground,
formed a painful contrast to the scene of poverty
and death.

The first six weeks of his career at the diggings had been most unsuccessful, and he had suffered as much from want as his unhappy wife. Then came a sudden change of fortune, and in two weeks more he was comparatively rich. He hastened immediately to Melbourne, and for a whole week had sought his family in vain. At length, on the preceding evening, he found them only in time to witness the last moments of his wife.

Sad as this history may appear, it is not so sad as many, many others; for often, instead of returning with gold, the digger is never heard of more.

In England many imagine that the principal labour at the diggings consists in stooping to pick up the lumps of gold which lie upon the ground at their feet, only waiting for some one to take possession of them. These people, when told of holes being dug in depths of from seven to forty feet before arriving at the desired metal, look upon such statements as so many myths, or fancy they are fabricated by the lucky gold-finders to deter too many others from coming to take a share of the precious spoil. There was a passenger on board the vessel which took me to Australia, who held some such opinions as these, and, although in other respects a sensible man, he used seriously to believe that every day that we were delayed by contrary winds he could have been picking up fifty or a hundred pounds worth of gold had he but been at the diggings. He went to Bendigo the third day after we landed, stayed there a fortnight, left it in disgust, and returned to England immediately—poorer than he had started.

This is not an isolated case. Young men of sanguine dispositions read the startling amounts of gold shipped from the colonies, they think of the "John Bull Nugget" and other similar prizes, turn a deaf ear when you speak of blanks, and determinately overlook the vast amount of labour which the gold diggings have consumed. Whenever I meet with this class of would-be emigrants, the remarks of an old digger, which I once over-

heard, recur to my mind. The conversation at
the time was turned upon the subject of the
many young men flocking from the "old
country" to the gold-fields, and their evident
unfitness for them. "Every young man before
paying his passage money," said he, "should
take a few days' spell at well-sinking in England ;
if he can stand that comfortably, the diggings
won't hurt him."

Many are sadly disappointed on arriving in
Victoria, at being unable to invest their capital
or savings in the purchase of about a hundred
acres of land, sufficient for a small farm. I
have referred to this subject before, but cannot
resist adding some facts which bear upon it.

By a return of the *land sales* of Victoria,
from 1837 to 1851, it appears that 380,000
acres of land were sold in the whole colony ;
and the sum realized by Government was
£700,000. In a return published in 1849, it
is stated that there were *three* persons who
each held singly more land in their own hands
than had been sold to all the rest of the colony

in fourteen years, for which they paid the sum of £30 each per annum. Yet, whilst £700,000 is realized by the *sale* of land, and not £100 a-year gained by *letting* three times the quantity, the Colonial Government persists in the latter course, in spite of the reiterated disapprobation of the colonists themselves ; and by one of the last gazettes of Governor La Trobe, he has ordered 681,700 acres, or 1,065 square miles, to be given over to the squatters. The result of this is, that many emigrants landing in Victoria are compelled to turn their steps towards the sister colony of Adelaide. There was a family who landed in Melbourne whilst I was there. It consisted of the parents, and several grown-up sons and daughters. The father had held a small tenant farm in England, and having saved a few hundreds, determined to invest it in Australian land. He brought out with him many agricultural implements, an iron house, &c. ; and on his arrival found, to his dismay, that no less than 640 acres of crown lands could be sold, at a time, at the

upset price of £1 an acre.   This was more
than his capital could afford, and they left for
Adelaide.   The expenses of getting his goods
to and from the ships, of storing them, of sup-
porting his family while in Melbourne, and of
paying their passage to Adelaide, amounted
almost to £100. Thus he lost nearly a fourth
of his capital, and Victoria a family who
would have made good colonists.

Much is done now-a-days to assist emigration, but far greater exertions are needed before
either the demand for labour in the colonies or
the over-supply of it in England can be exhausted.   Pass down the best streets of Melbourne: you see one or two good shops or
houses, and on either side an empty spot or a
mass of rubbish. The ground has been bought,
the plans for the proposed building are all
ready.   Then why not commence ?—there are
no workmen.   Bricks are wanted, and £15
a thousand is offered; carpenters are advertized
for at £8 a week; yet the building makes no
progress—there are no workmen.   Go down

towards the Yarra, and an unfinished church
will attract attention. Are funds wanting for
its completion? No. Thousands were subscribed in one day, and would be again were it
necessary; but that building, like every other,
is stopped for lack of workmen. In vain the
bishop himself published an appeal to the
various labourers required, offering the very
highest wages; others offered higher wages
still, and the church (up to the time I left
Victoria) remained unfinished. And yet, whilst
labour is so scarce, so needed in the colonies,
there are thousands in our own country *able
and willing to work*, whose lives here are one
of prolonged privation, whose eyes are never
gladdened by the sight of nature, who inhale
no purer atmosphere than the tainted air of the
dark courts and dismal cellars in which they
herd. Send them to the colonies—food and
pure air would at least be theirs, and much
misery would be turned into positive happiness.

I heard of a lady who every year sent out a

whole family from the poor but hard-working classes to the colonies (it was through one of the objects of her thoughtful benevolence that this annual act became known to me), and what happiness must it bring when she reflects on the heartfelt blessings that are showered upon her from the far-off land of Australia. Surely, among the rich and the influential, there are many who, out of the abundance of their wealth, could " go and do likewise."

THE END.

LONDON:
Printed by Schulze and Co., 13, Poland Street.